Problem-Solving Exercises in Green and Sustainable Chemistry

Problem-Solving Exercises in Green and Sustainable Chemistry

Albert S. Matlack
University of Delaware, Newark, USA

Edited by Andrew P. Dicks
University of Toronto, Ontario, Canada

CRC Press
Taylor & Francis Group
Boca Raton   London   New York

CRC Press is an imprint of the
Taylor & Francis Group, an **Informa** business

CRC Press
Taylor & Francis Group
6000 Broken Sound Parkway NW, Suite 300
Boca Raton, FL 33487-2742

© 2016 by Taylor & Francis Group, LLC
CRC Press is an imprint of Taylor & Francis Group, an Informa business

No claim to original U.S. Government works

Printed on acid-free paper
Version Date: 20150813

International Standard Book Number-13: 978-1-4822-5257-6 (Paperback)

Library of Congress Cataloging-in-Publication Data

Matlack, Albert S., 1923-
   Problem-solving exercises in green and sustainable chemistry / Albert S. Matlack and Andrew P. Dicks.
      pages cm
      "A CRC title."
      Includes bibliographical references and index.
      ISBN 978-1-4822-5257-6 (pbk. : alk. paper) 1. Green chemistry--Textbooks. 2. Chemistry--Textbooks. 3. Environmental management--Textbooks. I. Dicks, Andrew P. II. Title.

TP155.2.E58M39 2016
660--dc23   2015031037

Visit the Taylor & Francis Web site at
http://www.taylorandfrancis.com

and the CRC Press Web site at
http://www.crcpress.com

Contents

Foreword

As communication technology has advanced over the last few decades, people have been able to converse at lightning speed. Direct computer to computer e-mailing, phone to phone texting, and collective postings on various aggregate systems has provided an instantaneousness that helps to rapidly disseminate concepts and ideas across huge numbers of people. There are many reasons to celebrate these advances in communication. A more knowledgeable population of individuals should be better equipped to make decisions facing them. Unfortunately, for the obvious reasons of simplicity and efficiency, requirements governing the size and complexity of the information being transmitted have imposed a more abbreviated and simplified content. One could go on and on lamenting the sociological implications of this trend.

Abbreviated communication is relatively new to the human scene. Deep within our being, humans have been accustomed to and require the extended narrative. Even before the written word, storytelling around the campfire by the elders allowed the propagation and dissemination of critical information. Homer's epic poems *The Iliad* and *The Odyssey*, passed down through countless generations of oral history, are bursting with subtlety and nuance that provide philosophical introspection nearly three millennia later.

The field of green chemistry is not easily reduced into short, simple statements. There are levels of complexity that require the superimposition of several different and often opposing viewpoints relating impacts on human health and the environment. There is a somewhat fractal structure that is often revealed when delving into green chemistry implications. As more questions are answered, new questions arise whose answers give rise to new questions. Limiting communication to 140 characters not only omits the complexity of any topic but also gives the mistaken impression that it is easier or simpler than it really is. This creation of false expectations leads to obvious difficulties when people react to the partial information, often leading to unfortunate consequences.

This book offers a continuation of the human narrative that is green chemistry. Al Matlack has been part of the woven tapestry since the earliest days of the field. His industrial knowledge, insight, and genuine curiosity were uncontainable. Anyone attending any green chemistry symposium or conference grew to anticipate the inevitable question of every speaker who began with the phrase "Al Matlack, University of Delaware...." To some in the audience, Al's questions might have seemed to come from left field. Some, I suspect, grew impatient, not recognizing the significance of the questions he would ask. But his questions always connected seemingly disconnected concepts that were not only interesting but also significantly important to recognize. He would raise issues that could not be easily addressed at the end of a 45-minute presentation. Perhaps they were not suitable for abbreviated communication. But they were important.

The scientific society and the human society have lost the physical embodiment of Al Matlack. Andrew Dicks has combined his own insight and experience to help shape a book that provides a source of useful and well-developed information.

A quick scan of this book will not do it justice. This is not a collection of "sound bites." It is also not a complete comprehensive treatise of the field of green chemistry or sustainability. No individual or group of individuals could possibly be capable of representing an entire field. But what this book does offer is an excellent glimpse into how Al Matlack and Andrew Dicks view this complicated and nuanced area. For anyone trying to develop their own understanding of and relationship to green chemistry and sustainability, this book offers an opportunity to see through the viewpoint of deep experience and insight. With or without this book, Al Matlack will forever remain in our collective narrative.

John C. Warner
Warner Babcock Institute, LLC

Preface

This book is written with the intent of highlighting a method that makes our students want to come to class. A course covering aspects of green chemistry, industrial chemistry, and introductory polymer science was taught at the University of Delaware from 1995 to 2013. A semester of organic chemistry was a requirement for the course. A full-year prerequisite would certainly have been better, but students in chemical engineering generally do not take that much. Those enrolled were primarily seniors and first-year graduate students. Some of them worked an 8-hour day before arriving at class as they were often progressing toward a master's degree. Classes were three hours long and ran from 6:30 to 9:30 p.m. with a 10-minute break in the middle. They were conducted as interactive discussions rather than as formal lectures, as the latter can make it challenging for students to concentrate. The discussion format works only when the students have read the appropriate chapter in the course text (*Introduction to Green Chemistry*, 2nd Edition, Taylor & Francis) and other required outside reading before coming to class. The only straight lecture was when literature was covered since the text was written in 2010.

Just like anyone else, students can get tired after a day of studying or performing benchwork. However, they became excited to learn when the problem-based learning exercises described in this book were introduced for the last 30–45 minutes of class. The hope is that while less material can be covered this way, the student will remember more of it several years later. It also teaches them how to analyze real-world situations and how to be creative in solving problems. Many of these exercises are based on events that have actually taken place. In terms of organization, the class was split into small groups by letting them draw straws each week. Each group received a name, such as Paleocene, Pentacene, and Anthropocene, and the "problem of the week" to be discussed was presented. After they were no longer working on the problem, the solutions that the students devised were covered, with each group getting a turn. For each turn, a different spokesperson for the group spoke up, and this continued until the students exhausted their answers. The proposals were commented on and the spokesperson asked for further interpretation where necessary. It was emphasized to the class that every answer is of merit and is worthy of further consideration. A 10-minute quiz at the beginning of the next class might include a question regarding the problem-based learning exercise from the previous week.

The book blends some fundamental principles of green chemistry and sustainability from *Introduction to Green Chemistry* with a total of 50 problems discussed at the University of Delaware. Wherever possible, the problems are organized around appropriate content. As examples, Chapter 1 deals with issues surrounding substance toxicity, chemical accidents, and waste generation/disposal. In the associated end-of-chapter problems, students are placed in the real-world roles of a manufacturing plant chief executive officer, a chemistry department chair, and a chemical engineer. Chapter 4 concerns energy and the environment and the dilemma of a university architect who is designing new buildings expected to be more energy-efficient. Chapter 5 explores the development of new policies, with one problem focusing on the concern of too many cars on a university campus. Proposed solutions, examples, and comments are collected together in Chapter 7, along with relevant literature references. It is perhaps unsurprising to learn that students are often uncomfortable that there are several "correct" answers to many of the problems. It should be regularly emphasized to them that green chemistry is about decision-making and the weighing of many factors that are often conflicting.

Whether you are teaching green chemistry for the first time or are a seasoned instructor, we hope you find the problems described here both relevant and stimulating. Many thanks go to John Warner and Hal White for writing the foreword and introduction, respectively, and to Hilary LaFoe for her efforts in making this book come to fruition.

Albert S. Matlack
Department of Chemistry and Biochemistry
University of Delaware

Andrew P. Dicks
Department of Chemistry
University of Toronto

Introduction

Problem-based learning in principle is how we learn most of the time outside of school. We are confronted with a problem we have to solve. The way to proceed is not obvious. The problem may seem intractable or there may be many possible solutions, some better than others. One needs to explore possibilities and obtain sufficient information from different realms to make a decision on how to proceed given constraints and expectations. Traditional education is not structured this way. Typically information is first provided in lectures and texts, studied by students, and then applied to discipline-specific problems that often have a single correct answer.

In the early 1990s, a group of faculty in the sciences at the University of Delaware decided to adopt and adapt problem-based learning to their introductory undergraduate biology, chemistry, biochemistry, and physics courses. Instead of lecturing to students most of the time, we developed real-world problems that required the understanding of the basic science concepts that would have been presented in lecture. This was one of the first concerted efforts to teach using problem-based learning (PBL) at the undergraduate level. It was a few years after the arrival of PBL at the University of Delaware when Al Matlack, a retired chemist from Hercules, began teaching environmental chemistry and industrial chemistry as an adjunct professor at the university.

Sometime later, Al came to my office expressing his frustration with the poor performance of the students in his classes and wanting to learn about my experience with PBL in my biochemistry courses. As with many problems he encountered at work or in life, Al was looking for solutions. He had heard about PBL and wondered if it might be an alternative to his lecturing approach that did not seem to achieve his desired outcome of student learning and understanding. I suspect that was when Al began to think in terms of actively involving students in problem-solving situations in class. Because of Al's exceptionally broad interests and his long career as an industrial chemist, he had a wealth of experiences that could be transformed into real-world problems for students to work on.

Al's knowledge and experience was not limited to teaching in the classroom. He had broad interests in the environment that extended well beyond chemistry. As president of the Society of Natural History of Delaware, for many years he organized monthly field trips carefully selected to provide participants with first-hand exploration of environmental issues affecting society. In that context, I became a student of Al Matlack. He took our group to visit paper recycling operations where we saw what happened to paper products spared from the landfill. Our group visited neighborhoods built at different times to see how architecture changed with the advent of air conditioning and the effects that had on landscape plantings and neighborly interactions. With Al, I and others visited various storm water retention basins having different designs and learned about impervious surfaces, ground water restoration, and flood control. There were many other field trips relating to forest management and reforestation, invasive plant species and their impact, recovery of sturgeon populations in the Delaware Bay, the quaternary geology of the Delmarva

Peninsula showing evidence of permafrost, or removal of dams to permit spawning of anadromous fish. All could be considered problems requiring societal solutions, and many had chemical dimensions.

Some of those themes are echoed in the problems in this book. It is a gold mine of relevant problems pertaining to green chemistry and sustainability. The motivated student will find much to ponder here.

Harold B. White
Department of Chemistry and Biochemistry
University of Delaware

Author

To say that my father, **Albert S. Matlack**, was a chemist doesn't do him justice. Perhaps Chemist, capitalized in recognition of his passion for the subject, would be better. We have an essay he wrote at around the age of 13 stating his intention of becoming one. No alternatives were mentioned. His chemical career started with the Manhattan project during World War II. Following that, he was an organic chemist for 43 years at Hercules Incorporated in Wilmington, Delaware, retiring at the age of 70 only when forced to do so. Despite his age at the time, he had not yet had enough chemistry and promptly volunteered to teach at the University of Delaware, which he did until only months before his death at the age of 90 in 2013. That also wasn't enough to satisfy his appetite for chemistry, so at the same time, over the course of more than 10 years, he wrote a (big, long) book, *Introduction to Green Chemistry.* Then he wrote this one, finishing it—by my calculation—only days before his death. In the months before he died, he and I were approaching chemical journals, successfully, to find him a place where he could write a regular column commenting critically on recent chemical developments. Now, when I am reading the chemical literature and come across something particularly cool, among my thoughts is "It's sad that Dad will never get to know this."

You might find such prolonged focus on a technical topic a bit frightening and fear that he was narrow and boring in person. He was not. He was warm, outgoing, and friendly and a relentlessly and passionately constructive man (and a great husband and father). For me, his attitude can be summarized in something he once said while we were discussing the study of history as an occupation: "Some people are interested in the past. I'm interested in the future."

My father did not join an ongoing, established green chemistry field. Rather, he pre-dated it by decades. He was an environmentalist before the term was coined. Green chemistry combined his two passions: chemistry and the environment. He could clearly see the damage humans are doing to the planet, and he focused his almost unbelievable energy and patience on the problem. He would encourage you to do the same because the problem is acute, getting worse, and of a magnitude that dwarfs anything we have ever dealt with before. Politics aside, much of it comes down to chemistry. We need new energy sources, less pollution, better batteries, renewable starting materials...the list is long. We also need well-informed, passionate people who understand chemistry and can use it creatively to solve these problems. If you contribute to their solution, you will really have accomplished something. As the bumper stickers say: "There is no Planet B."

Go for it, and good luck.

Kent E.S. Matlack, PhD

Editor

Andrew P. Dicks, PhD, joined the University of Toronto Chemistry Department in 1997. After undergraduate and graduate study in the United Kingdom, he became an organic chemistry sessional lecturer in 1999, and he was hired as part of the university teaching-stream faculty two years later. He has research interests in undergraduate laboratory instruction that involve designing novel and stimulating experiments, particularly those that showcase green chemistry principles. This work has lead to over 45 peer-reviewed publications in the chemical education literature. Following promotion in 2006, he became Associate Chair of Undergraduate Studies for two years and developed an ongoing interest in improving the student experience in his department.

Dr. Dicks has won several pedagogical awards, including the University of Toronto President's Teaching Award, the Chemical Institute of Canada Award for Chemical Education, a 2011 American Chemical Society-Committee on Environmental Improvement Award for Incorporating Sustainability into Chemistry Education and a 2015 Canadian Green Chemistry and Engineering Award. He has additionally edited a book as a resource for teaching green chemistry (*Green Organic Chemistry in Lecture and Laboratory*, CRC Press). In 2014 he was co-chair of the 23rd IUPAC International Conference on Chemistry Education, which was held in Canada for the first time since 1989.

Following the passing of Albert Matlack in 2013, Dr. Dicks assumed editorship of *Problem-Solving Exercises in Green and Sustainable Chemistry* in order to ensure the issues discussed in this book became available to the broader chemistry community.

1 Toxicity, Accidents, and Chemical Waste

1.1 GENERAL BACKGROUND

This chapter considers what is toxic, what is waste, why accidents occur, and how to reduce all of these.[1] In the glorious days of the 1950s and 1960s, chemists envisioned chemistry as the solution to a host of society's needs. Indeed, they created many of the things we use today and take for granted. The discovery of Ziegler–Natta catalysis of stereospecific polymerization alone resulted in major new polymers. The chemical industry grew by leaps and bounds until it employed about 1,027,000 workers in the United States in 1998.[2] By 2007, this number had dropped to 872,200. Some may remember the DuPont slogan, "Better things for better living through chemistry." In the Sputnik era, the scientist was a hero. At the same time, doctors aided by new chemistry and antibiotics felt that infectious diseases had been conquered.

Unfortunately, amid the numerous success stories were some adverse outcomes that chemists had not foreseen.

Thalidomide (Schematic 1.1) was used to treat nausea in pregnant women from the late 1950s to 1962. It was withdrawn from the market after 8000 children in 46 countries were born with birth defects.[3] The compound has other uses as a drug as long as it is not given to pregnant women. In Brazil, it is used to treat leprosy. Unfortunately, some doctors there have not taken the warning seriously enough and several dozen deformed births have occurred.[4] The U.S. Food and Drug Administration (FDA) has approved its use for treating painful inflammation of leprosy.[5] It also inhibits human immunodeficiency virus (HIV) and can prevent the weight loss that often accompanies the acquired immunodeficiency syndrome (AIDS). Celgene is using it as a lead compound for an anti-inflammatory drug and is looking for analogs with reduced side effects.[6] The analog overleaf (Schematic 1.2) is 400–500 times as active as thalidomide. Revlimid (Schematic 1.3) is approved for treating multiple myeloma.[7]

Today, there is often public suspicion toward scientists.[8] Some picture a mad chemist with his stinks and smells. There is a notion among some people "that science is boring, conservative, close-minded, devoid of mystery, and a negative force in society."[9] Chemophobia has increased. Many people think that chemicals are bad and "all natural" is better, even though a number of them do not know what a chemical is. There is a feeling that scientists should be more responsible for the influence of their work on society. Liability suits have proliferated in the United States. This has caused at least three companies to declare bankruptcy: Johns Manville for asbestos in 1982, A. H. Robins for its "Dalkon Shield" contraceptive device in 1985, and Dow Corning for silicone breast implants in 1995.[10] Doctors used to be the respected pillars of their communities. Today they are subjects of malpractice suits, some of which only serve to increase the cost of health care. Medical implant research is threatened by the unwillingness of companies

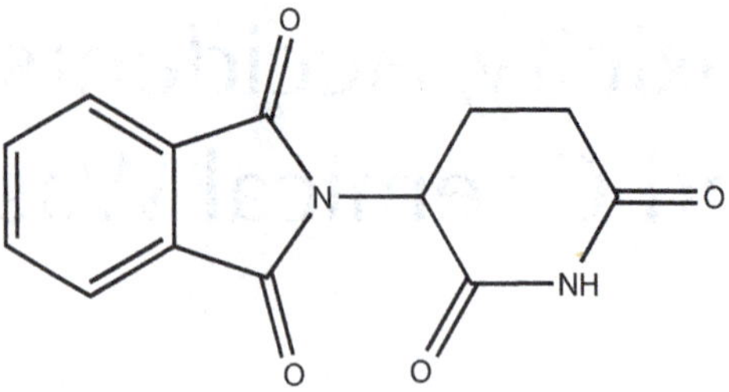

SCHEMATIC 1.1   Thaliomide

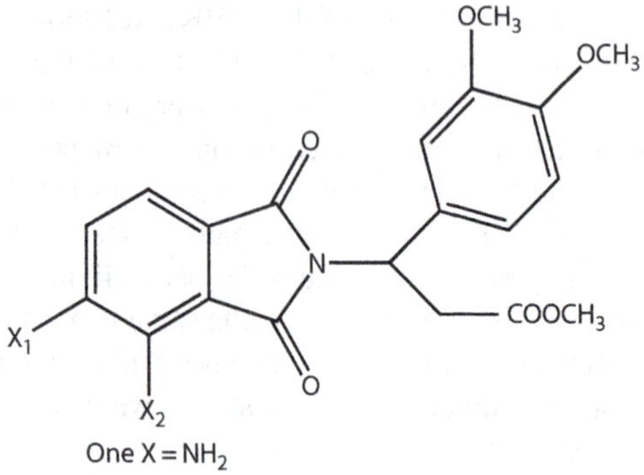

One X = NH$_2$

SCHEMATIC 1.2   Analog of thalidomide

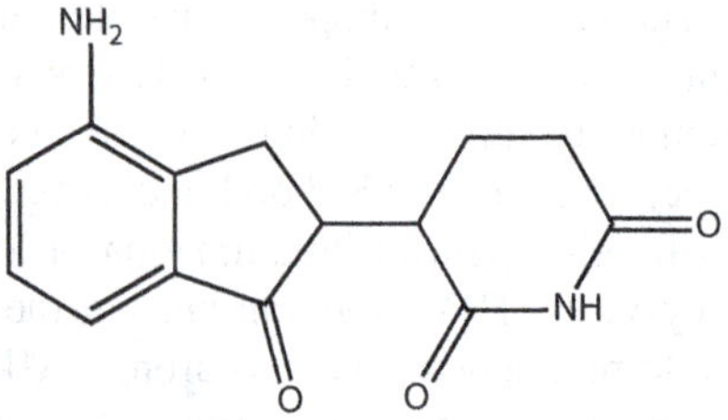

SCHEMATIC 1.3   Revlimid

such as DuPont and Dow Corning to sell plastics for the devices to implant companies.[11] The chemical companies fear liability suits. Not long ago drug companies became so concerned about lawsuits on childhood vaccines that many were no longer willing to make them. Now that the U.S. Congress has passed legislation limiting the liability, vaccine research is again moving forward. The lawsuits had not stimulated research into vaccines with fewer side effects, but instead had caused companies to leave the market.

1.2 TOXICITY OF CHEMICALS IN THE ENVIRONMENT

The public's perception of toxicity and risk often differs from that found by scientific testing.[12] The idea that "natural"[13] is better than "chemical" is overly simplistic. Many chemicals that are found in nature are extremely potent biologically. Mycotoxins are among these.[14] Aflatoxins (Schematic 1.4) were discovered when

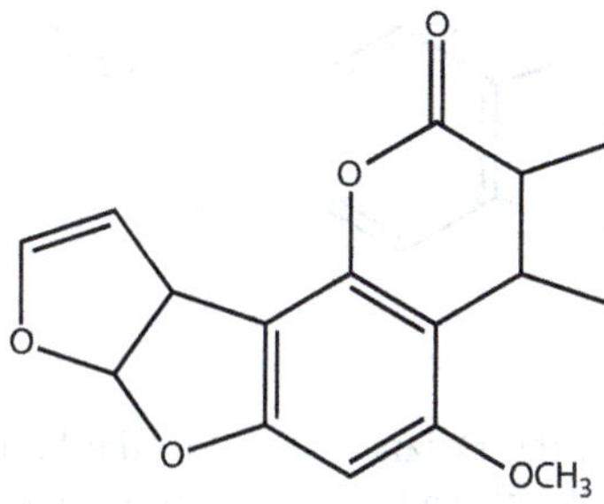

SCHEMATIC 1.4 Aflatoxin B₁

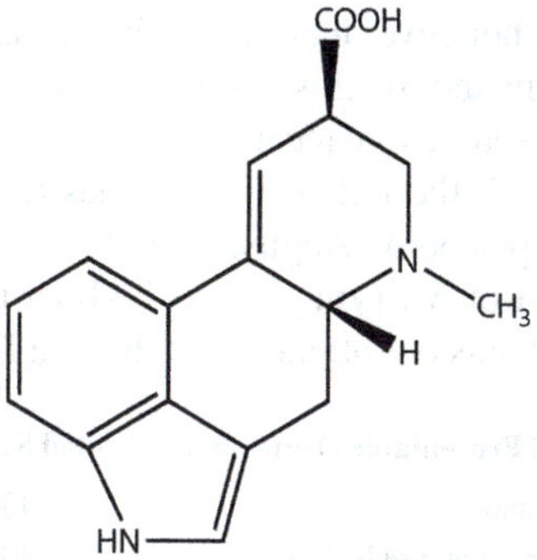

SCHEMATIC 1.5 Lysergic acid

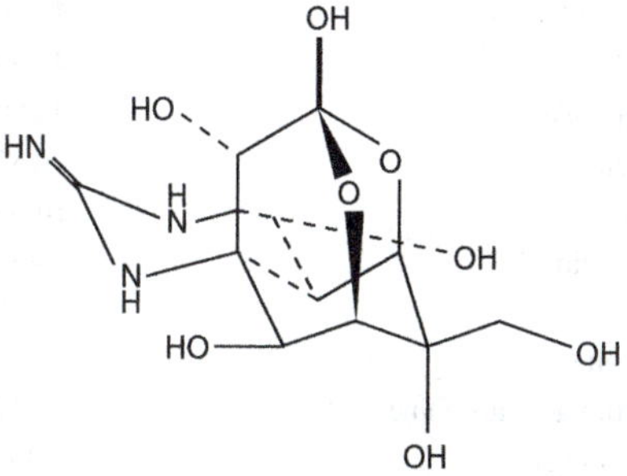

SCHEMATIC 1.6 Tetrodotoxin

turkeys fed moldy groundnut (peanut) meal became ill and died. They are among the most potent carcinogens known. Vikings went berserk after eating derivatives of lysergic acid (Schematic 1.5) made by the ergot fungus growing on rye.

Some *Amanita* mushrooms are notorious for the poisons that they contain. A Japanese fish delicacy of globefish or other fishes may contain potent poisons (such as tetrodotoxin; Schematic 1.6)[15] if improperly prepared.[16] Tainted fish cause death in 6–24 hours in 60% of those who consume it. They die from paralysis of the lungs. Oysters may also contain poisons acquired from their diet. The extract of the roots of the sassafras tree (*Sassafras albidum*) (Schematic 1.7) is used to flavor the soft drink "root beer" but contains the carcinogen safrole, which must be removed before use.

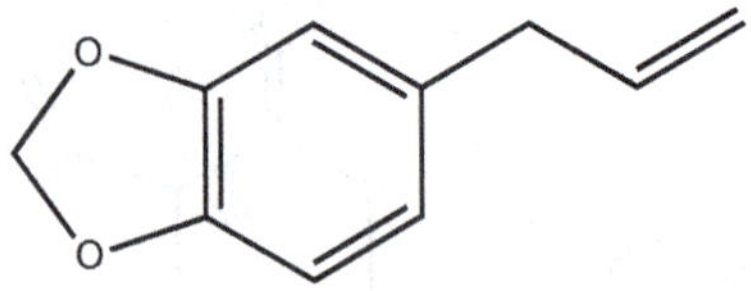

SCHEMATIC 1.7   Safrole

The U.S. Congress added the Delaney clause to the Food, Drug, and Cosmetic Act in 1958.[17] The clause reads: "No additive shall be deemed to be safe if it is found to induce cancer when ingested by man or animal, or if it is found, after tests which are appropriate for the evaluation of the safety of food additives, to induce cancer in man or animal." It does not cover natural carcinogens in foods or environmental carcinogens, such as chlorinated dioxins and PCBs. Some Americans feel that food additives are an important source of cancer, and that the Delaney clause should have been retained. Bruce Ames,[18] the father of the Ames test for mutagens, disagrees. The Delaney clause was repealed on August 3, 1996.[19]

Ames feels that, instead of worrying so much about the last traces of contaminants in foods, we should focus our attention on the real killers.

Annual Preventable Deaths in the United States[20]

Active smoking	430,700
Overweight and sedentary	400,000
Alcoholic beverages	100,000
Passive smoking	53,000
Auto accidents	43,300
AIDS	37,500
Homicides	34,000
Suicides	30,575
Falls	14,900
Drowning	4400
Fires	3200
Cocaine	4202
Heroin and morphine	4175
Bee stings	3300
Radon, to nonsmokers	2,500
Lightning	82
Recalcitrant farm animals	20
Dog mauling	17
Snakebite	12

The World Health Organization (WHO) estimates that the number of smoking deaths worldwide is about 3 million/year.[21] Cigarette use is increasing among American college students, despite these statistics.[22] Indoor radon contributes to about 12% of the lung cancer deaths in the United States each year.[23] Infectious diseases cause 37% of all deaths worldwide.[24] Many of these could be prevented by improved sanitation. There are 3 million pesticide poisonings, including 220,000 fatalities and 750,000 chronic illnesses, in the world each year. Smoke from cooking with wood fires kills 4 million children in the world each year.

It is estimated that perhaps 80% of cancers are environmental in origin and related to lifestyle. There is "convincing" evidence of a connection between excess weight and cancers of the colon, rectum, esophagus, pancreas, kidney, and breast in postmenopausal women.[25] (There are genetic factors, now being studied by the techniques of molecular biology, that predispose some groups to heightened risk, e.g., breast cancer in women.[26]) In addition to cancers caused by tobacco and ethanol, there are those caused by being overweight,[27] too much sun, smoked foods, foods preserved with a lot of salt, and viruses (for cancers of the liver and cervix). Consumption of the blue-green alga, *Microcystis*, has increased liver cancer in China.[28] Many species of cyanobacteria produce the neurotoxin (Schematic 1.8).[29]

Perhaps the biggest killer is sodium chloride, a compound necessary for life, which plays a role in the regulation of body fluids and blood pressure.[30] It raises the blood pressure of many of the 65 million Americans with hypertension, increasing the risk of osteoporosis, heart attack, and stroke.[31] The U.S. National Academy of Sciences suggests limiting the consumption of sodium chloride to 4 g/day (1.5 g of sodium).[32] This means cutting back on processed foods (source of 80% of the total), such as soups, frozen dinners, salted snacks, ham, soy sauce (18% salt), ready-to-eat breakfast cereals, and others. The Dietary Approaches to Stop Hypertension (DASH) diet, which reduces salt, has lots of fruits and vegetables, and minimizes saturated fat from meat and full-fat milk, is as effective as drugs in lowering blood pressure in many cases. However, this may not be the whole story. If there is an adequate intake of calcium, magnesium, and potassium, together with fruits and vegetables in a low-fat diet, the sodium may not need to be reduced, as shown in a 1997 study.[33] The DASH diet also reduces the risk of heart disease by 24% and of stroke by 18%.[34] People who do not smoke, are active physically, drink alcohol in moderation, and eat at least five servings of fruits and vegetables a day live 14 years longer on average than those who do none of these.[35] In the United States, snacks and soft drinks have tended to supplant nutrient-rich foods, such as fruits, vegetables, and milk. Not eating fruits and vegetables poses a greater cancer risk than traces of pesticides in foods.[36] Fruits and vegetables often contain natural antioxidants,[37] such as the resveratrol (Schematic 1.9) found in grapes. Resveratrol inhibits tumor initiation, promotion, and progression.[38] Sirtus Pharmaceuticals was founded in 2004 to develop derivatives of resveratrol for age-related diseases.[39]

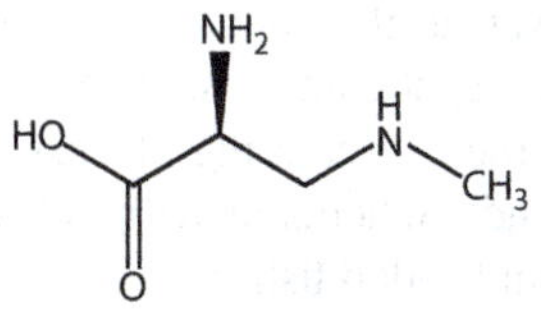

SCHEMATIC 1.8   Neurotoxin from cyanobacteria

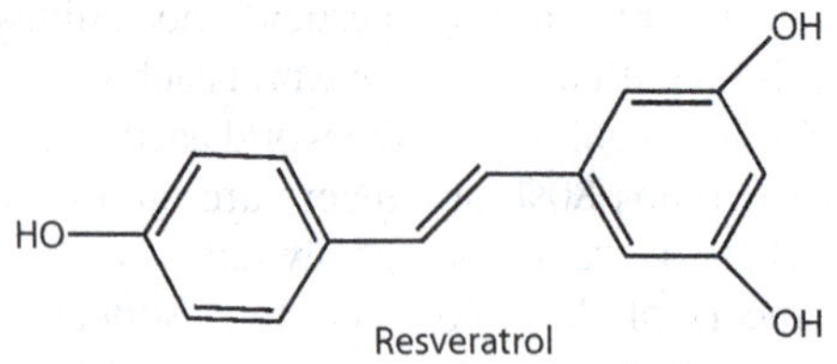

SCHEMATIC 1.9   Resveratrol

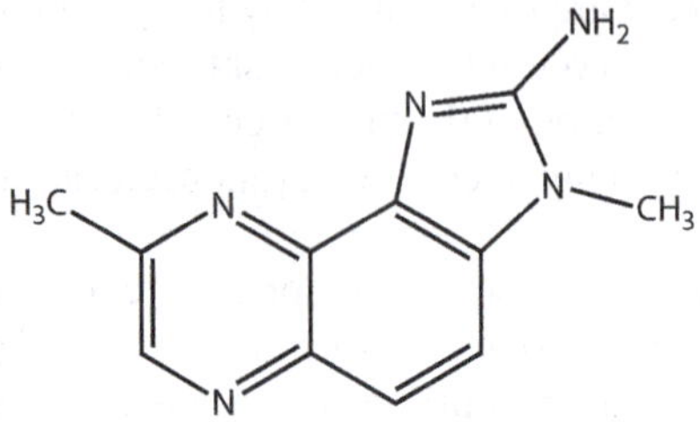

SCHEMATIC 1.10   Heterocyclic amine mutagen found in fried beef

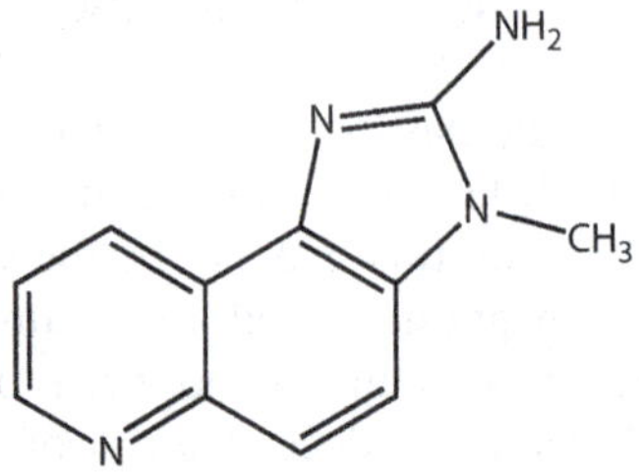

SCHEMATIC 1.11   Heterocyclic amine mutagen found in broiled fish

Antioxidants have also reduced atherosclerotic heart disease.[40] Thus, foods contain many protective substances, as well as some antinutrients, such as enzyme inhibitors and natural toxins.[41]

A U.S. National Research Council report concludes that natural and synthetic carcinogens are present in human foods at such low levels that they pose little threat.[42] It points out that consuming too many calories as fat, protein, carbohydrates, or ethanol is far more likely to cause cancer than consuming the synthetic or natural chemicals in the diet. However, it also mentions several natural substances linked to increased cancer risk: heterocyclic amines, such as nitrosamines, aflatoxins, and other mycotoxins, formed in the overcooking of meat.[43] Typical of the heterocyclic amine mutagens are compounds Schematic 1.10 and Schematic 1.11, the first from fried beef and the second from broiled fish.[44]

Deep frying with soybean, sunflower, and corn oil allows air oxidation of linoleates to highly toxic (2E)-4-hydroxy-2-nonenal (Schematic 1.12), which has been linked to Parkinson's and Alzheimer's diseases.[45]

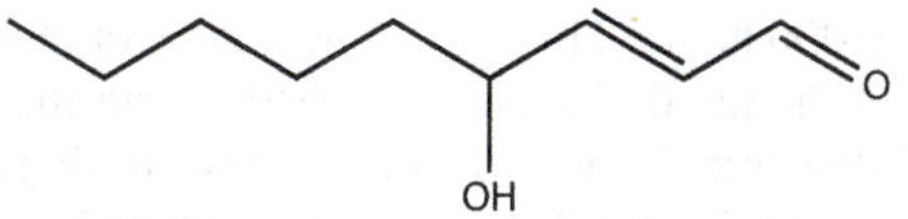

SCHEMATIC 1.12   (2E)-4-hydroxy-2-nonenal

Two reviews cover the incidence of cancer and its prevention by diet and other means.[46] Cancer treatments have had little effect on the death rates, so prevention is the key.[47]

Being sedentary is a risk factor for diseases such as heart attack and late-onset diabetes. One-third of adult Americans are obese, perhaps as much the result of cheap gasoline as the plentiful supply of food.[48] Obesity-related complications result in 400,000 premature deaths in the United States each year. This is less of a problem in most other countries. For example, the incidence of obesity in the United Kingdom is 20%.

Prevention of disease is underused.[49] Needle exchange programs could prevent 17,000 AIDS infections in the United States each year. Vaccines are not used enough. For example, only 61% of the people in Massachusetts are fully vaccinated. Only 15%–30% of elderly, immunocompromised persons, and those with pulmonary or cardiac conditions have been vaccinated against pneumonia. A 1998 study, in New Jersey and Quebec, of patients older than 65 who had been prescribed cholesterol-lowering drugs found that, on average, the prescription went unfilled 40% of the year. Good drug compliance lowered the cholesterol level by 39%, whereas poor compliance lowered it by only 11%.[50]

1.3   ACCIDENTS WITH CHEMICALS

Chemists take pride in their ability to tame dangerous chemicals in an effort to make the things that society needs. In fact, some companies seek business by advertising their ability to do custom syntheses with such chemicals.[51] Aerojet Fine Chemicals offers syntheses with azides and vigorous oxidations.[52] Carbolabs offers custom syntheses with phosgene, fluorinating agents, and nitration. Custom syntheses with phosgene are also offered by PPG Industries, Hatco, Rhone-Poulenc, and SPNE. Pressure Chemical, Dynamit Nobel, and Rutgers suggest that they can do the hazardous reactions for others.[53] A hazardous reagent may be attractive for fine chemical syntheses if it gives a cleaner product with less waste or saves two or three steps. It may also be used because it is the traditional way of doing the job. Methods for screening unknown reactions for hazards have been summarized.[54]

Chemistry is a relatively safe occupation. (Underground coal mining is one of the most dangerous in the United States. A total of 47 coal miners lost their lives in 1995 from surface and underground mining.[55]) In the United States in 1996, the nonfatal injury and illness rate for chemical manufacturing was 4.8:100 full-time workers, compared with 10.6 for all of manufacturing. There were 34 deaths in the chemical industry, about 5% of those for all manufacturing.[56] The injury rate of the chemical industry in the United Kingdom fell to an all-time low in 1997,

0.37 accidents per 100,000 hours. This was in the middle of those for manufacturing industries, worse than the textile industry, but better than the food, beverage, and tobacco industries.[57] However, despite countless safety meetings and inspections and safety prizes, accidents still happen.[58] In chemistry, just as in airline safety, some of the accidents can be quite dramatic. There were 23,000 accidents with toxic chemicals in the United States in 1993–1995 (i.e., 7670/year compared with 6900/year in 1988–1992). The accidents in 1993–1995 included 60 deaths and the evacuation of 41,000 people. Statistics for the United States compiled by the Chemical Safety and Hazard Investigation Board reveal an average of 1380 chemical accidents resulting in death, injury, or evacuation each year for the 10 years before 1999.[59] Each year, these accidents caused an average of 226 deaths and 2000 injuries. About 60,000 chemical accidents are reported annually in the United States. The American Chemistry Council in the United States reported 793 fires, explosions, and chemical releases in 1995.[60] American Chemistry Council members reported about 2000 accidents in 2003, about the same as a decade before.[61] About 50% of global chemical incidents are in the United States and 30% in the EU.[62] The United Kingdom is second only to the United States in recorded incidents.

Knowing the cause should help eliminate accidents in the future. Then engineering steps may be taken to produce a fail-safe system. These may involve additional alarms, interlocks (such as turning off microwaves before the oven door can be opened), automatic shutoffs if any leaks occur, and secondary vessels that would contain a spill. A special sump contained a leak of nerve gas when an O-ring failed at an incinerator at Tooele, Utah.[63] Underground storage tanks (made of noncorroding materials) may be fitted with a catchment basin around the fill pipe, automatic shutoff devices to prevent overfilling, and a double-wall complete with an interstitial monitor.[64] Clearly, such methods can work, but they have not reduced the overall incidence of accidents, as shown in the foregoing data. The best solution will be to satisfy society's needs with a minimum of hazardous chemicals.[65] A few of the many accidents will be discussed in the following to show how and why they occurred, together with some green approaches that could eliminate them.

The explosion of the nuclear reactor at Chernobyl (spelling changed recently to Chornobyl) in the Ukraine on April 26, 1986 sent radioactive material as far away as Sweden.[66] The current death toll is 65. There has been a huge increase in childhood thyroid cancer, with cases as far as 500 km away.[67] (U.S. bomb tests have also increased the incidence of thyroid cancers in the western United States.[68]) There is a 30-km exclusion zone around the plant where no one is allowed to live. This was created by the evacuation of 135,000 people.[69] The accident is said to have happened "because of combination of the physical characteristics of the reactor, the design of the control rods, human error and management shortcomings in the design, and implementation of the safety experiment."

The realization of the seriousness of global warming has led to many proposals for more nuclear plants.[70] Electricity need not be generated by nuclear power. Generating it from fossil fuels contributes to global warming. Producing it from renewable sources, such as wind, wave power, hydropower, geothermal, and solar energy, does not. Sweden has voted to phase out nuclear energy. The German government has agreed to phase out the country's 19 nuclear reactors.[71] It has been estimated that offshore wind power sources could produce electricity 40% more cheaply than the nuclear power stations planned for Japan.[72] Energy conservation can help a great deal in reducing the amount of energy needed.

The flammable gases used by the petrochemical industry have been involved in many accidents.[73] A fire and explosion following a leak of ethylene and isobutane from a pipeline at a Phillips plant in Pasadena, Texas, in 1989, killed 23 people and injured 130.[74] The Occupational Health and Safety Administration (OSHA) fined the company US$4 million. Accidents of this type can happen anywhere in the world where petrochemical industries are located. There have been explosions at an ethylene plant in Beijing, China,[75] at a Shell air separation plant in Malaysia,[76] at a Shell ethylene and propylene plant in Deer Park, Texas,[77] at a Shell propylene plant in Norco, Louisiana,[78] at a BASF plant in Ludwigshafen, Germany, that used pentane to blow polystyrene,[79] and at a Texaco refinery at Milford Haven, England.[80] The problems at the BASF plant could have been avoided by blowing the polystyrene with nitrogen or carbon dioxide instead of pentane. Two weeks before this incident at the BASF plant, four people were injured by a fire from a benzene leak.[81] BASF had two other accidents in October 1995, a polypropylene fire in Wilton, England, and spraying of a heat transfer fluid over the plant and adjacent town in Ludwigshafen, Germany.[82] There was an explosion in the hydrogenation area of the company's 1,4-butanediol plant in Geismar, Louisiana, on April 15, 1997,[83] which resulted from internal corrosion of a hydrogen line. This corrosion might have been detected by periodic nondestructive testing with ultrasound. The explosion and fires at the Texaco refinery have been attributed to making modifications in the plant but not training people on how to use the modified equipment, having too many alarms (2040 in the plant), insufficient inspection of the corrosion of the equipment, not learning from experience, and reduced operator staffing.[84] Investigation of an explosion and fire at Shell Chemical's Belpre, Ohio, thermoplastic elastomer plant revealed that at the time of the accident, roughly seven times the normal amount of butadiene had inadvertently been added to the reactor. Alarms indicated that the reactor had been overcharged, but interlocks were manually overridden to initiate the transfer of raw materials into the reactor vessel, contrary to established procedures.[85] The federal government fined the company US$3 million for the various citations in relation to the accident.[86]

These examples show that even though companies have a great deal of experience in handling hazardous materials, accidents can still occur. This includes toxic gases as well. A leak in the hydrogen cyanide unit at a Rohm & Haas, Deer Park, Texas plant sent 32 workers to the hospital.[87] Presumably, the hydrogen cyanide was being used to react with acetone in the synthesis of methyl methacrylate. An alternative route that does not use hydrogen cyanide is available.[88] Isobutylene is oxidized catalytically to methacrolein and then to methacrylic acid, which is esterified with methanol to give methyl methacrylate. The methacrolein can also be made by the hydroformylation of propyne, although this does involve the use of toxic carbon monoxide and flammable hydrogen.[89] These processes also eliminate the ammonium bisulfate waste from the process using hydrogen cyanide. A leak of hydrogen fluoride at a Marathon Petroleum plant in Texas City, Texas, sent 140 people to the hospital for observation and treatment of inflamed eyes and lungs and caused the evacuation of 3000 more.[90] Replacement of the hydrogen fluoride with a nonvolatile solid acid would eliminate such a problem. Many companies in the United Kingdom do not store hazardous materials correctly, which has resulted in some fires and explosions.[91]

More than 44 million Americans live or work near places that pose risks from the storage or use of dangerous industrial chemicals.[92] The cost of accidents may be more than just a monetary one to the company. A fire and explosion occurred on July 4, 1993, at a Sumitomo Chemical plant in Niihama, Japan, that made over half of all the epoxy-encapsulation resin for semiconductor chips used in the entire world. Cutting off the supply would have been a serious inconvenience to the customers. The company took the unusual step of letting other companies use its technology until it could rebuild its own plant, so that a supply crisis never developed. The company still supplied 50% of the world's requirements for that resin in 1999.[93]

Why do these accidents continue to happen? One critic says, "Hourly workers struggle to maintain production in the face of disabled or ignored alarms, undocumented and often uncontrollable bypasses of established components, operating levels that exceed design limits, postponed and severely reduced turnaround maintenance and increasing maintenance on 'hot' units by untrained, temporary non-union contract workers."[94] Another source mentions "institutional realities that undercut corporate safety goals, such as incentives that promote safety violations in the interest of short-term profitability, shielding upper management from 'bad news' and turnover of management staff."[95] A third says that "Many chemical plant disasters have been precipitated by an unplanned change in process, a change in equipment or a change in personnel."[96]

1.4   WASTE AND ITS MINIMIZATION

Nearly everything made in the laboratory ends up as waste. After the materials are made, characterized, and tested, they may be stored for a while, but eventually they are discarded. In schools, the trend is to run experiments on a much smaller scale, which means that less material has to be purchased and less waste results.[97] (Even less material will be used if chemistry-on-a-chip becomes commonplace.[98]) Two things limit how small a scale industrial chemists can use. One is the relatively large amount of a polymer needed for the fabrication of molded pieces for testing physical properties. There is a need to develop smaller scale tests that will give data that are just as good. The second factor is the tendency of salesmen to be generous in offering samples for testing by potential customers. A lot of the samples received by the potential customer may never be used.

Some industrial wastes result because it is cheaper to buy new material than to reclaim used material. Some catalysts fall into this category. The 1996 American Chemical Society National Chemical Technician Award went to a technician at Eastman Chemical who set up a program for recovering cobalt, copper, and nickel from spent catalysts for use by the steel industry.[99] This process for avoiding landfill disposal gave Eastman significant savings. Some waste metal salts can be put in fertilizer as trace elements that are essential for plant growth. However, this practice has been abused in some cases by putting in toxic waste (e.g., some that contain dioxins and heavy metals).[100] Even a waste as cheap as sodium chloride can be converted back to the sodium hydroxide and hydrochloric acid that it may have come from, by electrodialysis using bipolar membranes.[101] Waste acid can be recovered by vacuum distillation in equipment made of fluoropolymers.[102]

Improved housekeeping can often lead to reduced emissions and waste.[103] The leaky valves and seals can be fixed or they can be replaced with new designs that minimize emissions.[104] These include diaphragm valves, double mechanical seals with interstitial liquids, magnetic drives, better valve packings, filled fabric seals for floating roof tanks, and so on. Older plants may have some of the worst problems.[105] The U.S. Environmental Protection Agency (EPA) fined Eastman Kodak US$8 million for organic solvents leaking from the 31 miles of industrial sewers at its Rochester, New York, plant. This emphasizes the need for regular inspections and preventive maintenance, which in the long run is the cheapest method. Vessels with smooth interiors lined with non- or low-stick poly(tetrafluoroethylene) can be selected for batch tanks that require frequent cleaning. These might be cleaned with high-pressure water jets instead of solvents. Perhaps a vessel can be dedicated to a single product, instead of several, so that it does not have to be cleaned as often. If a product requires several rinses, the last one can be used as the first one for a new lot of product. Volatile organic compounds can be loaded with dip tubes instead of splash loading. Exxon (now Exxon-Mobil) has used such methods to cut emissions of volatile organic compounds by 50% since 1990.[106] Install automatic high-level shutoffs on tanks. Use wooden pallets over again, instead of considering them throwaway items. The current ethic should be "reduce," "reuse," and "recycle" in that order.

A waste is not a waste if it can be reused. For example, one steel manufacturer drops pickle liquor down a 100-ft-tall tower at 1200°F to recover iron oxide for magnetic oxide and hydrogen chloride for use again in pickle liquor.[107] The pomace left over from processing pears and kiwis can be dried and used to increase the dietary fiber in other foods.[108] Food-processing wastes and wastes from biocatalytic processes often become feed for animals. A refinery stream of ethane, methane, butane, and propane, which was formerly flared as waste, will be processed to recover propane for conversion to propylene and then to polypropylene.[109] Organic chemical wastes may end up as fuel for the site's power plant or for a cement kiln, but more valuable uses would be preferable. Waste exchanges are being set up. One company's waste may be another's raw material. For example, calcium sulfate from flue gas desulfurization in Denmark and Japan ends up in drywall for houses. If the waste exchange merely pairs up an acid and a base so that the two can be neutralized, rather than reclaimed, the result is waste salts. Admittedly, these are probably not as toxic as the starting acid and base, but they still have to be disposed of somewhere. Several general references on waste minimization and pollution prevention are available.[110]

The hydroformylation[111] of propene to form butyraldehyde invariably produces some isobutyraldehyde at the same time (Schematic 1.13).[112] One of the best processes uses a water-soluble rhodium phosphine complex to produce 94.5% of the former and 4.5% of the latter.[113] The products form a separate layer that is separated from the water. Rhodium is expensive, so it is important to lose as little as possible.

SCHEMATIC 1.13 Synthesis of butyraldehyde from propene

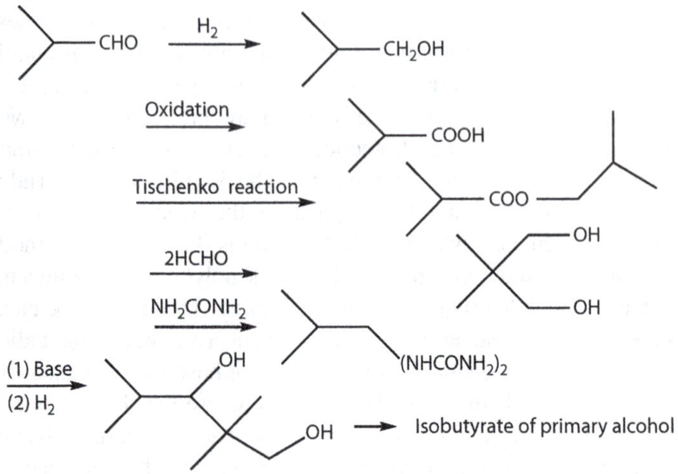

SCHEMATIC 1.14 Useful products from isobutyraldehyde

In 10 years of operation by Rhone-Poulenc-Ruhrchemie, 2 million metric tons of butyraldehyde have been made with the loss of only 2 kg of rhodium. The process is 10% cheaper than the usual one. Higher olefins are not soluble enough in water to work well in the process. The process does work for omega-alkenecarboxylic acids such as 10-undecenoic acid, where a 97:3 normal/iso compound is obtained in 99% conversion.[114] For higher alkenecarboxylic acids, a phase-transfer catalyst, such as dodecyltrimethylammonium bromide, must be used. However, this lowers the normal/iso ratio.

Over the years a variety of uses have been found for isobutyraldehyde by Eastman Chemical and others.[115] It is converted to isobutyl alcohol, neopentyl glycol, isobutyl acetate, isobutyric acid, isobutylidenediurea, methylisoamyl ketone, and various hydrogenation and esterification products (Schematic 1.14).

1.5 CONCLUSIONS

The challenge is to reduce the incidence and severity of accidents, waste, the toxicity of chemicals, and the amount of energy used, while still providing the goods that society needs. Several provocative papers suggest some ways of doing this.[116] The key is in the preparation of more sophisticated catalysts. Thus, solid acids may be able to replace the risky hydrogen fluoride and sulfuric acid used in alkylation reactions in the refining of oil. Zeolites offer the promise of higher yields through size and shape selectivity. With the proper catalysts, oxidations with air and hydrogen peroxide may replace heavy metal-containing oxidants. Enantioselective catalysis may

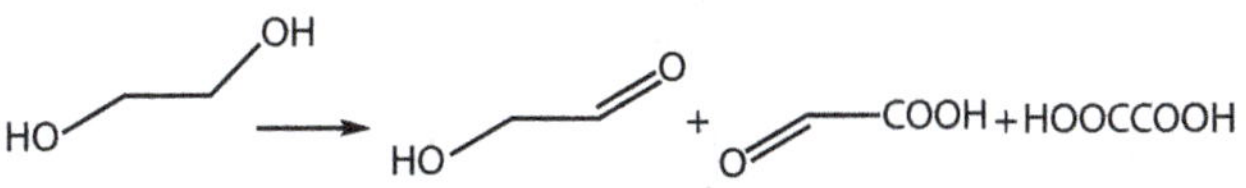

SCHEMATIC 1.15   Biotransformation of ethylene glycol

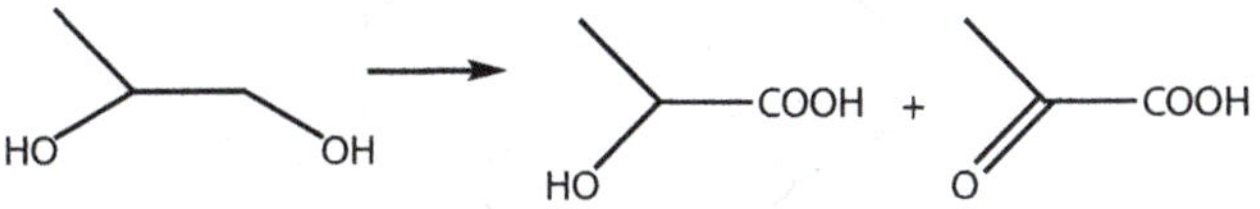

SCHEMATIC 1.16   Biotransformation of propylene glycol

allow the preparation of the biologically active optical isomer without the unwanted one. It may be possible to run the reaction in water at or near room temperature using biocatalysis instead of in a solvent or at high temperature. Some processes yield more by-product salts than the desired product. Sheldon[117] recommends a salt-free diet by improved catalytic methods.

Bodor[118] has suggested the design of biologically safer chemicals through retrometabolic design. For example, the ethylene glycol used widely as an antifreeze in cars might be replaced with less hazardous propylene glycol. The former is converted by the body to glycolaldehyde, glyoxylic acid, and oxalic acid (Schematic 1.15), whereas the latter gives the normal body metabolites lactic acid and pyruvic acid (Schematic 1.16).[119]

A lethal dose of ethylene glycol for a man is 1.4 mL/kg. The problem is that its sweet taste makes it attractive to children and pets. An alternative approach is to add a bittering agent to it. The estimated lethal dose of propylene glycol for a man is 7 mL/kg.

The U.S. National Science and Technology Council has laid out a research and development strategy for toxic substances and waste.[120] Hirschhorn[121] has suggested ways of achieving prosperity without pollution.

Propylene used to be converted to isotactic polypropylene with titanium tetrachloride and diethylaluminum chloride in a hydrocarbon solvent.[122] The atactic polypropylene obtained by evaporation of the solvent after filtration of the desired isotactic polymer was of little value, some going into adhesives. An acidic deashing step was necessary to remove residual titanium, aluminum, and chloride from the polymer, the metal-containing residues ending up in a landfill. This process was supplanted by high mileage catalysts during which the titanium chloride was supported on magnesium chloride.[123] These were activated by triethylaluminum in the presence of ligands that enhance the stereoselectivity of the catalysts. The result was a product that required no deashing and no removal of atactic polymer. In the next step in the evolution, the solvent was eliminated by polymerization in the gas phase or in liquid propylene. By the proper choice of catalyst, the polymer can be obtained in large enough granules so that the older practice of extruding molten polymer to form a strand that was chopped into "molding powder" is no longer necessary. The field is still evolving. Metallocene single-site catalysts[124] allow greater control of the product and have led to new products. Ethylene-α-olefin copolymers can be made from ethylene alone,

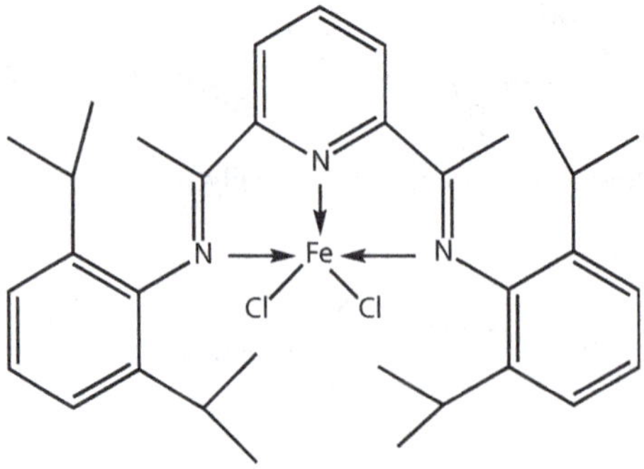

SCHEMATIC 1.17   Catalyst for copolymerization

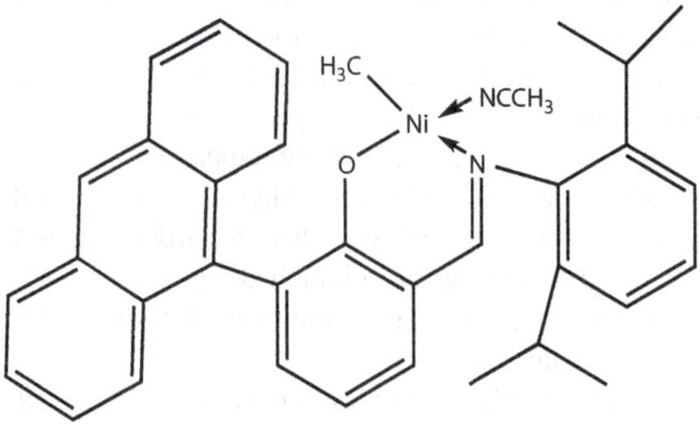

SCHEMATIC 1.18   Water-active polymerization catalyst

the α-olefin being made in situ.[125] A typical catalyst is shown in Schematic 1.17. It is activated by methyl alumoxane. Methods of preparing syndiotactic polypropylene in a practical way are now available. A new polypropylene made with such catalysts may be able to supply the properties now found only in plasticized polyvinyl chloride. All but the last type of catalyst show little tolerance to air, moisture, and polar groups. The catalyst in Schematic 1.18 is active in the presence of ethers, ketones, esters, and water.[126] Ethylene can even be polymerized in water by one of the palladium diimine catalysts.[127] The initial polymerization probably forms a shell of polyethylene that protects the catalyst from the water. To make the products even greener, consumer use must be reduced, the lower level of reuse and recycle must be raised, and a renewable source of the propylene, rather than petroleum, must be used. Propylene could be made by reduction and then dehydration of acetone from fermentation.

Hirschhorn[128] feels that "An environmentally driven industrial revolution is beginning." Brain Rushton, President of the American Chemical Society in 1995, said that "We will gradually eliminate environmentally unsound processes and practices from our industry. We will build a better environment to work and live in. We will keep scientists and engineers both employed and at the cutting edge of technology as they serve the competitive needs of the nation. Last but by no means least, we will create a better image for chemistry and the profession."[129]

Lastly, Gro Harlem Brundtland, formex head of the WHO and former prime minister of Norway and the Secretary General of the World Commission on Environment and Development, has noted that "The obstacles to sustainability are not mainly technical. They are social, institutional and political."

PROBLEMS

1.1 Pollution Prevention at an Isocyanates Plant

You are the chief executive officer at an isocyanates manufacturing plant based in North America. Isocyanates are used in the manufacture of pesticides (methyl isocyanate was the compound responsible for the 1984 Bhopal disaster in India, which is a highly toxic lachrymator). With the help of a nongovernmental organization, a study by company engineers has identified several opportunities to prevent pollution at one of your major production plants. The savings would be $1 million per year, with payback periods potentially ranging from 15 months to 5 years. The newly designed processes would eliminate 500,000 lb. of waste per year, and allow the expensive hazardous waste incinerator to be decommissioned. In addition to the savings in dollars, it could drastically help to improve public relations.

Considering the significant up-front capital expenses involved, should you implement the plan? If so, how would you implement it?

1.2 Curious Polar Bears

Some polar bears on Svalbard in the high arctic above Norway were found to have sexual organs of both sexes. This looks like a case of endocrine disruption. How would you find out what compound or compounds were involved? How did they get to this remote place? Pesticides are likely, but there is no gardening on Svalbard. What could be done about it, if anything?

1.3 After an Industrial Accident

You are the on-site manager of a production plant that uses a lot of chlorine gas. Recently, a storage tank ruptured due to an accident, which released about 500 lb. of chlorine. Fortunately, weather conditions prevented the gas from settling over nearby towns. However, if a similar incident happens there may be serious repercussions, such as necessary evacuation of houses. What can you do (both mechanically and chemically) to ensure that there will be no reoccurrence of this incident?

1.4 A Mighty Safety Dilemma

You are the chairman of the Chemistry Department at a state university with 20,000 students. Your department has had an exemplary safety record for many years, even winning awards for it. However, something has gone wrong. There have been three incidents in the last two weeks. Although there have been no injuries, you know that the difference between an incident and an injury, or even a fatality, may depend on where the person is standing at the time. Since these incidents, the parade of administrators through your office and the building has been steady. This is more attention than your department has received in years. You have also met the emergency team of the state environmental agency, the local fire department, and the OSHA inspectors.

Your primary concern is that no student gets hurt when undertaking experimental work. You remember a woman who lost an eye when the solution that she was boiling in a test tube bumped and hit her in the face. Three other concerns are (1) will the adverse publicity reduce the number of students taking chemistry?; (2) will the alumni stop donating to the department?; and (3) will the US$1 million bill for the cleanup ruin the department's budget for the year? You want to be viewed as a strong leader who takes action and not as a wishy-washy one who is indecisive. Should you stop all laboratory work for a month of intensive safety training? The first incident involved a student who added ether to toluene, which was heated above the boiling point of the ether. The fire occurred when the vapors ignited on the hot plate. The second incident involved the explosion of a stoppered glass bottle to which concentrated nitric acid and organic waste had been added. Apparently, the student involved did not realize that nitrogen oxides form when organic matter is oxidized by nitric acid. The third incident was a 3:00 a.m. explosion and fire involving a Diels-Alder reaction being carried out in a sealed glass vessel. This last incident set off the fire sprinkler system, which flooded both the third floor of the wing and the floors beneath it. Who was responsible for these incidents? While students were present in each case, their professors should have given them better training. You realize that it is your responsibility to see that the professors do their jobs. It is up to you to devise a carrot-stick approach that will prevent future problems of this type.

Which, if any, of the following should you do?

1. Call the incidents "human errors." Tell the students that chemistry can be dangerous and that they should be more careful.
2. Chastise the students and the professors for being stupid.
3. Get a psychologist from the university Psychology Department to see if the students and their professors were under undue stress at the time.
4. Have the students fill out detailed safety reports on the incidents, then go over these with their professors, then with the professor in charge of safety, and finally with you.
5. Put the students on probation or suspend them until they complete a rigorous course in safety training.
6. Encourage the students to transfer to another department.
7. Demote the professors or delay their getting tenure.
8. See if the professors are spending all their time writing grant proposals so that they do not devote enough time to their students.
9. Hire an outside safety consultant to assess the department.
10. Solicit help from safety engineers at local companies.
11. Appoint a different professor to be in charge of departmental safety.
12. Adopt the methods used by industry, for example, unannounced safety inspections, safety prizes, safety slogans in prominent places, etc.
13. Post tables of flash points of common solvents at every desk.

14. Prohibit the use of ether and other solvents with low flash points.
15. Require training in the use of alternative reaction media with higher flash points.
16. Prohibit the use of concentrated nitric acid and/or require that all acids be neutralized prior to disposal.
17. Put a copy of the American Chemical Society manual on disposal of wastes on every desk.
18. Require that all waste containers be vented to a fumehood.
19. Provide more training before students are allowed to work on their own.
20. Limit the scale of reactions to no more than one gram. Require that all students and professors take a course in microscale chemistry. Purchase some microscale glassware for the undergraduate and graduate laboratories.
21. Require all students and faculty to take a course in inherently safer chemistry, that is, green chemistry.
22. Require a permit for all unattended overnight reactions to be issued after a review with a safety engineer.
23. Require that all unattended overnight reactions be run first by differential scanning calorimetry to see how exothermic they are and to see what pressures develop.
24. Insist that all reactions in glass vessels that may develop more than 5 lb. of pressure be encased in heavy metal cans (with small vents) that will contain glass if it breaks. A plastic shield should be used in front of this and the reaction should be conducted in a fumehood.
25. Build a new facility for reactions to be done under pressure, where work is done in steel autoclaves remotely behind several inches of concrete wall. Or, farm out the work to a nearby facility that has such equipment.
26. Provide every lab with a "Thermo-Watch" temperature controller.
27. Have all students and faculty take a training course in conducting reactions under pressure. This should include handling of high-pressure gas cylinders.
28. Use combinatorial chemistry to find a suitable catalyst system for a Diels-Alder reaction.
29. Hunt for alternative, less hazardous media for the Diels-Alder reaction.
30. Initiate a big brother or big sister system in the labs where every new student is mentored by a more experienced graduate student for the first two years.
31. Provide for an automatic shutdown of the sprinkler system when sensors indicate that a fire is out.
32. Rebuild the wing with a floor drain in every laboratory.
33. Do not allow carpets on the floors of offices.
34. Do not allow any files, file cabinets, or stacks of papers within one foot of the floor. Anything placed closer than a foot from the floor must be something that will not be damaged by water.
35. Solicit ideas from all faculty members, including those in the Department of Chemical Engineering.

1.5 To Burn or Not to Burn?

You are in charge of a state agency that regulates air and water emissions. A large company has applied for a permit to incinerate about 12,000 t of hazardous waste each year at its research center, which is located in a residential area. About three-quarters of the waste will be brought in from other sites of the company. Although the company has announced a goal of zero waste, substantial amounts of chlorinated solvents will be burned. You wonder why the company is still using these since they cause liver damage and cancer. Why is it not switching to processes that use less solvent and less harmful solvents? Why is it not practicing more recovery and reuse of solvents?

Trial burns have shown emissions of As, Be, Cd, and Cr(VI) to be within EPA limits. No data has been presented for Hg, which is the metal most likely to come out of the stack. The company does have a program to collect batteries and fluorescent lamps separately. However, the compliance rate of employees is unknown to you. Medical incinerators are the largest source of dangerous chlorodioxins, and the new incinerator could produce these. The company has no data on these from trial burns. The amounts of particulates will be low, but there has been no check for harmful particles under 2.5 μm, the size that a person's breathing system has problems with. Since the 75 ft. stack of the incinerator is located in a narrow valley, it might be possible for the untrapped 3 lb./day to linger in a mass of stagnant air in the valley on a day when there is no wind. Could this be enough to annoy or harm local residents? The company has a good record, and most of what it proposes is credible. How should you proceed?

1.6 "Delacid": A Versatile Catalyst

A local inventor has approached your university for funding for a revolutionary acid catalyst. It is insoluble under aqueous conditions and in all organic solvents. It can be made available in each of liquid, solid, and film forms. The material can be used whenever Brønsted and Lewis acid catalysts are needed. It is capable of replacing hydrofluoric acid, sulfuric acid, boron trifluoride, aluminum chloride, and so on, and is completely stable to air and water. Recovery for reuse is simple, and catalyst life is measured in years. Spent catalyst can be used as a bulking agent in hamburgers. Catalyst cost in large volumes is estimated to be as little as $1.00/lb.

You strongly suspect that this appears too good to be true. However, you feel obliged to see what the catalyst actually is before some state legislator inquires about it. How can you characterize the few grams of powder that the inventor has left with you, and how can you decide what it is actually good for?

1.7 The Flask Broke

A 5-L round-bottomed flask broke when a solvent was being dried over a sodium/potassium alloy. The student was burned, but recovered and graduated from the school. How could this accident have been prevented?

1.8 The Tremendous Problem of Climate Change

Climate change is certainly here. It is caused by emission of carbon dioxide, a greenhouse gas that has reached 400 ppm in the atmosphere. Devise ways to synthesize the polymers that modern society relies on that will release no carbon dioxide.

1.9   The Chemist Talks to the Chemical Engineer

Pretend that you are a chemical engineer in charge of scaling up an industrial process. Your chemist colleague is proud of the overall yield of 90% that he has achieved in a multistep methodology. The starting material cost is only $1.00/g when purchased from Sigma-Aldrich. The first step is an oxidation with acidic potassium dichromate to produce a carboxylic acid. This is converted to a methyl ester with diazomethane. The ester is subsequently reduced by lithium aluminum hydride in diethyl ether. The resulting alcohol is then protected by reaction with dihydropyran, with the protecting group being removed after the next step.

A further reaction step produced a 100% yield when a 5% solution in dioxane/2-methoxyethanol was left for 48 hours at room temperature using a mercuric acetate catalyst. The following step gave the best yields when run over a 24-hour period at -78 °C. A final step worked best at a pressure of 7 kbar. The desired product was purified by distillation at 0.01 Torr using a 100-plate column. You point out to the chemist that he should have talked to you sooner. What has he done wrong, and why?

1.10   Inherently Safer Chemistry

Deaths and injuries in the chemical industry have not declined in the past decade. How many of these does it take to make a ton of chemicals? Environmentalists and trade unions have pushed for inherently safer chemistry that follows the principles of green chemistry. Industry objects violently, saying that "We don't need government officials telling us how to run our business." What does this really mean? Surely they do not condone deaths in the workplace. What are they afraid of? The Toxics Reduction Act of Massachusetts has worked well for the past 25 years. Why is industry complying with it while objecting to inherently safer chemistry? What is its secret?

1.11   A Strange Malady

There are no longer any young peregrine falcons, eagles, ospreys, and pelicans. As older birds die off, the species will become extinct. The only eagles will be in sports teams and Boy Scout patrols. Florida pelicans attract tourists, so its tourism revenue will eventually decline. Is this analogous to the canary in the coal mine? How can the cause be determined? What can be done about the problem?

1.12   The Mysterious Case of the Disappearing Filter Paper

You are a professor at a large university in North America. One of your students reported the disappearance of the filter paper when he tried to filter a reaction mixture containing an ionic liquid. You felt that something was wrong, so you had him repeat the process. It still gave the same result. You had the student repeat the experiment as you watched. The filter paper disappeared again. You called the manufacturer to be sure that they had not supplied substandard filter paper. They sent a new batch that gave exactly the same result. Should you use inorganic filter paper or another medium for filtration purposes, such as Celite? Should you call Sherlock Holmes for help? Since the effect is definitely real, what can you do to use it to your advantage?

REFERENCES

1. I.T. Horvath and P.T. Anastas, eds, *Chem. Rev.*, 2007, *107*, 2167 (issue on green chemistry).
2. Anon., *Chem. Eng. News*, May 18, 1998, 19; November 12, 2007.
3. (a) T. Stephens and R. Brynner, *Dark Remedy: The Impact of Thalidomide and its Revival as a Vital Medicine*, Perseus, Cambridge, MA, 2001; (b) T. Stephens, *Chem. Br.*, 2001, *37*(11), 38.
4. Anon., *Harvard Health Lett.*, 1998, *23*(4), 4.
5. R. Hoffmann, *Chem. Eng. News*, Apr. 29, 1996.
6. Anon., *Chem. Ind. (London)*, 1998, 591.
7. Anon., *Chem. Eng. News*, Dec. 18, 2006, 27.
8. J. Schummer, B. Bensaude-Vincent, and B. van Tiggelen, eds, *The Public Image of Chemistry*, World Scientific Publishing, Singapore, 2007.
9. (a) Anon., *Chem. Ind.*, 1996, 396; (b) C. Djerassi, *Science*, 1996, *272*, 1858.
10. D.R. Hofstadter, *Science*, 1998, *281*, 512.
11. M. Reisch, *Chem. Eng. News*, May 22, 1995, 6.
12. R.F. Service, *Science*, 1994, *266*, 726.
13. (a) R.S. Stricoff, *Handbook of Laboratory Health and Safety*, 2nd ed., Wiley, New York, 1995; (b) National Research Council, *Prudent Practices in the Laboratory: Handing and Disposal of Chemicals*, National Academy of Sciences Press, Washington, DC, 1995; (c) R. Rawls, *Chem. Eng. News*, Aug. 14, 1995, 4; (d) R.E. Lenga, *The Sigma-Aldrich Library of Chemical Safety Data*, 2nd ed., Sigma Aldrich Corp, Milwaukee, WI, 1988; (e) N.I. Sax and R.A. Lewis, *Dangerous Properties of Industrial Materials*, 7th ed., van Nostrand–Reinhold, New York, 1989; (f) D.V. Sweet, R.L. Sweet, Sr., eds, *Registry of Toxic Effects of Chemicals Substances*, Diane Publishers, Upland, PA, 1994; (g) C. Maltoni and I.J. Selikoff, Living in a chemical world—occupational and environmental significance of industrial carcinogens, *Ann. NY Acad. Sci.*, 1988, *534*, whole issue; (h) J. Lynch, *Kirk–Othmer Encyclopedia of Chemical Technology*, 4th ed., Wiley, New York, 1995, *14*, 199; (i) B. Ballantyne, *Kirk–Othmer Encyclopedia of Polymer Science and Engineering*, 2nd ed., 1989, *16*, 878; (j) R.P. Pohanish and S.A. Greene, *Hazardous Materials Handbook*, Wiley, New York, 1996; (k) R.J. Lewis, Sr., *Hazardous Chemicals Desk Reference*, 4th ed., Wiley, New York, 1996; (l) G. Schuurmann, B. Market, *Ecotoxicology, Ecological Fundamentals, Chemical Exposure and Biological Effects*, Wiley, New York, 1997; (m) P. Calow, *Handbook of Ecotoxicology*, Blackwell Scientific, London, 1993; (n) M. Richards, *Environmental Xenobiotics*, Taylor & Francis, London, 1996.
14. B.M. Jacobson, *Chem. Eng. News*, Jan. 11, 1999, 2.
15. R.D. Coker, *Chem. Ind. (London)*, 1995, 260.
16. (a) S. Budavaria, M.J. O'Neil, A. Smith, P.E. Heckelman, and J.F. Kinneary, eds., *Merck Index*, 12th ed., Merck & Co., Whitehouse Station, NJ, 1996, 1578; (b) Anon., *Tufts University Diet and Nutrition Letter*, 1996, *14*(6), 8.
17. D. Hanson, *Chem. Eng. News*, Sept. 12, 1994, 16.
18. B.N. Ames and L.S. Gold, *Angew. Chem. Int. Ed.*, 1990, *29*, 1197; *258*, 261.
19. Anon., *Amicus J.*, 1996, *18*(3), 4.
20. (a) Anon., *ASH Smoking Health Rev.*, July–Aug. 1998, 7; (b) T. Levin, *On Earth*, 2008, *30*(2), 44; (c) Anon., *Harvard Health Lett.*, 2001, *26*(8), 6; (d) Anon., *World Watch*, 2002, *15*(6), 40; (e) Anon., *University of California Berkeley Wellness Lett.*, 2004, *20*(9), 1.
21. Anon., *New Sci.*, Apr. 25, 1998, 19.
22. H. Wechsler, N.A. Rigotti, J. Gledhill-Hoyt, and H. Lee, *JAMA*, 1998, *280*, 1673.
23. Anon., *Environ. Sci. Technol.*, 1998, *32*, 213A.

24. (a) D. Pimentel, M. Tort, L. D'Anna, A. Krawic, J. Berger, J. Rossman, F. Mugo, N. Doon, M. Shriberg, E. Howard, S. Lee, and J. Talbot, Ecology of increasing disease, population growth and environmental degradation, *BioScience*, 1998, *48*(10), 817; (b) Anon., *1998–1999 World Resources: A Guide to the Global Environmental*, World Resources Institute, Washington, DC; (c) Anon., *Chem. Eng. News*, May 11, 1998, 21.

25. www.dietandcancerreport.org.

26. F.P. Perera, *Science*, 1997, *278*, 1068.

27. (a) Anon., *Johns Hopkins Medical Lett., Health After 50*, 1996, *8*(6), 1; (b) M. Nestle, *Science*, 2003, *299*, 781, 846–859 (special section on obesity); (c) F as in Fat: How Obesity Policies Are Failing in America 2007, www.healthyamericans.org/reports/obesity2007; (d) C. Runyan, *World Watch*, 2001, *14*(2), 11.

28. R. Schuhmacher, *Chem. Ind(London)*, July 17, 2006, 24.

29. Anon., *Chem. Eng. News*, Apr. 11, 2005, 26.

30. Anon., University California Berkeley Wellness Lett., Sept. 1998, 14(12), 6.

31. (a) Anon., *Johns Hopkins Medical Lett., Health After 50*, 1996, *8*(2), 6; (b) Anon., *Tufts University Diet and Nutrition Lett.*, 1996, *14*(4), 1; (c) 1996, *14*(5), 4–6; (d) Anon., *Tufts Health and Nutrition Lett.*, Feb. 2006, special supplement on salt.

32. (a) Anon., Duke Medicine Health News, 2008, 14(6), 12; (b) Anon., University of California Berkeley Wellness Lett., 2007, 23(6), 1.

33. (a) G. Taubes, *Science*, 1998, *281*, 898; (b) D.A. McCarron, *Science*, 1998, *281*, 933.

34. (a) Anon., *Tufts Health & Nutrition Lett.*, 2008, *26*, 1; (b) N.R. Cook, J. Cutler, E. Obarzanek, J.E. Buring, K.M. Rexrode, S.K. Kumanyika, L.J. Appel, and P.K. Whelton, *Br. Med. J.*, 2007, *334*, 885.

35. Anon., University of California Berkeley Wellness Lett., 2008, 24(8), 1.

36. (a) Anon., *Environ. Sci. Technol.*, 1998, *32*, 81A; (b) Anon., *Chem. Eng. News*, Nov. 24, 1997, 60.

37. (a) S.J. Risch and C.-T. Ho, eds, *Spices: Flavor Chemistry and Antioxidant Properties.* A.C.S. Symp. 660, Washington, DC, 1997; (b) J.T. Kumpulainen and J.T. Salonen, *Natural Antioxidants and Anticarcinogens in Nutrition, Health and Disease*, Special Pub. 240, Royal Society of Chemistry, Cambridge, 1999.

38. M. Jang, L. Cai, G.O. Udeani, K.V. Slowing, C.F. Thomas, C.W.W. Beecher, H.H.S. Fong, N.R. Farnsworth, A.D. Kinghorn, R.G. Mehti, R.C. Moon, and J.M. Pezzuto, *Science*, 1997, *275*, 218.

39. Y. Bhattacharjee, *Science*, 2008, *320*, 593.

40. M.N. Diaz, B. Frei, J.A. Vita, and J.F. Keaney, Jr., *N. Engl. J. Med.*, 1997, *337*, 408.

41. F. Shahidi, Ed., *Antinutrients and Phytochemicals in Foods*, A.C.S. Symp. 662, Washington, DC, 1997.

42. National Research Council, *Carcinogens and Anticarcinogens in the Human Diet*, National Academy of Science Press, Washington, DC, 1996.

43. (a) Anon., *Chem. Ind. (London)*, 1996, 159; (b) J. Long, *Chem. Eng. News*, Feb. 26, 1996, 7; (c) Anon., *Environ. Sci. Technol.*, 1996, *30*(5), 199A.

44. H. Kasai, Z. Yamaizumi, T. Shiomi, S. Yokoyama, T. Miyazawa, K. Wakabayashi, M. Nagao, T. Sugimura, and S. Nishimura, *Chem. Lett.*, 1981, 485.

45. Anon., *Chem. Eng. News*, May 9, 2005, 35.

46. (a) C.M. Williams, *Chem. Ind. (London)*, 1993, 280; (b) W. Watson and B. Gloding, *Chem. Br.*, 1998, *34*(7), 45; (c) www.dietandcancerreport.org; (d) J. Ali, *Chem. Ind. (London)*, May 19, 2003, 15.

47. J.C. Bailar, III and H.L. Gornik, *N. Engl. J. Med.*, 1997, *336*, 1569.

48. (a) I. Wickelgren, *Science*, 1998, *280*, 1364; (b) G. Taubes, *Science*, 1998, *280*, 1367; (c) L.A. Campfield, F.J. Smith, and P. Burn, *Science*, 1998, *280*, 1383.

49. B.R. Jasny and F.E. Bloom, *Science*, 1998, *280*, 1507.

50. Anon., University California Berkeley Wellness Lett., 1999, 15(5), 8.

51. K.J. Watkins, *Chem. Eng. News*, Aug. 13, 2001, 17.
52. (a) Advertisement, *Chem. Eng. News*, Jan. 11, 1999 (inside back cover); (b) S.C. Stinson, *Chem. Eng. News*, July 13, 1998, 57, 71; *Chem. Eng. News*, Jan. 27, 2003, 1.
53. Anon., *Chem. Eng. News*, July 10, 2000, 70; *Chem. Eng. News*, Nov. 22, 1999, 33; *Chem. Eng. News*, Apr. 21, 2003, 30; *Chem. Eng. News*, Feb. 14, 2000, 94.
54. G. Amery, *Aldrichchim. Acta*, 2001, *34*(2), 61.
55. K. Snyder, U.S. Mine Safety and Health Administration, 1997.
56. A.M. Thayer, *Chem. Eng. News*, Apr. 27, 1998, 15.
57. Anon., *Chem. Br.*, 1998, *34*(8), 13.
58. F.P. Lees, *Loss Prevention in the Process Industries*, Butterworth–Heinemann, Oxford, 1996.
59. (a) J. Johnson, *Chem. Eng. News*, Mar. 15, 1999, 12; (b) Anon., *Chem. Eng. News*, Nov. 2, 2009, 22.
60. G. Parkinson, *Chem. Eng.*, 1997, *104*(1), 21.
61. Anon., *Chem. Eng. News*, Apr. 19, 2004, 29.
62. J. Cheftel, *Chem. Ind. (London)*, Mar. 7, 2005, 12.
63. *Chem. Eng. News*, Dec. 1998, 23.
64. W. Stellmach, *Chem. Eng. Prog.*, 1998, *94*(8), 71.
65. (a) F.R. Spellman and N.E. Whiting, *Safety Engineering: Principles and Practices*, Government Institutes, Rockville, MD, 1999; (b) T.A. Kletz, *Process Safety: A Handbook of Inherently Safer Design*, 2nd ed., Taylor & Francis, Philadelphia, PA, 1996; (c) D.A. Crowl, ed., *Inherently Safer Chemical Processes*, A.I.Ch.E., New York, 1996.
66. (a) M. Freemantle, *Chem. Eng. News*, Apr. 29, 1996, 18; (b) C. Hohenemser, F. Warner, B. Segerstahl, and V.M. Novikov, *Environment*, 1996, *38*(3), 3; (c) D. Hanson, *Chem. Eng. News*, Sept. 12, 2005, 11; (d) K. Charman, *World Watch*, 2006, *19*(4), 12, 19; (e) J. Bohannon, *Science*, 2005, *309*, 1663; (f) R. Stone, *Science*, 2006, *312*, 180; (g) R.K. Chesser and R.J. Baker, *Am. Sci.*, 2006, *94*, 542.
67. (a) M. Balter. *Science*, 1996, *270*, 1758; (b) J. Webb. *New Sci.*, Apr. 1, 1995, 7.
68. J. Johnson, *Chem. Eng. News*, Aug. 17, 1997, 10.
69. R. Stone, *Science*, 1998, *281*, 623.
70. (a) D. Whitford, *Fortune*, Aug. 6, 207, 42; (b) M. Freemantle, *Chem. Eng. News*, Sept. 13, 2004, 31; (c) E. Marshall, D. Clery, G. Yidong, and D. Normile, *Science*, 2005, *309*, 1168, 1172, 1177; (d) M. Jacoby, *Chem. Eng. News*, Aug. 24, 2009, 14; (e) A. Kadak, *Technol. Rev.*, 2009, *112*(2), 11.
71. Anon., *Amicus J.*, 1999, *21*(1), 16.
72. Anon., *Chem. Eng. News.*, Jan. 18, 33.
73. S.L. Wilkinson, *Chem. Eng. News*, Nov. 9, 1998, 71.
74. A.M. Thayer, *Chem. Eng. News*, Aug. 27, 1998, 15.
75. Anon., *Chem. Eng. News*, July 7, 1997, 22.
76. (a) Anon., *Chem. Eng. News*, June 29, 1998, 18; (b) P.M. Morse, *Chem. Eng. News*, Mar. 23, 1998, 17.
77. Anon., *Chem. Eng. News*, June 30, 1997, 19.
78. G. Peaff, *Chem. Eng. News*, Jan. 16, 1995, 16.
79. Anon., *Chem. Eng.*, 1998, *105*(10), 48.
80. Anon., *Chem. Eng. Prog.*, 1998, *94*(4), 86.
81. Anon., *Chem. Eng. News*, Aug. 17, 1998, 11.
82. Anon., *Chem. Eng. News*, Oct. 16, 1995, 9.
83. Anon., *Chem. Eng. News*, Apr. 28, 1997, 13; May 26, 1997, 12.
84. Anon., *The Explosion and Fires at the Texaco Refinery*, Milford Haven, 24 July, 1994, HSE Books, Sudbury, Suffolk, U.K.
85. Anon., *Chem. Eng. News*, Nov. 7, 1994, 9.
86. Anon., *Chem. Eng. News*, Dec. 5, 1994, 11.

87. Anon., *Chem. Eng. News*, Oct. 24, 1994, 10.

88. D. Arntz, *Catal. Today*, 1993, *18*, 173.

89. R.A. Sheldon, *Chemtech*, 1994, *24*(3), 38.

90. W. Worthy, *Chem. Eng. News*, Nov. 9, 1997, 6.

91. G. Ondrey, *Chem. Eng.*, 1998, *105*(6), 29.

92. (a) *Nowhere To Hide*, National Environmental Law Center, Boston, 1995; (b) *Chem. Ind. (London)*, 1995, 677.

93. J.-F. Tremblay, *Chem. Eng. News*, Jan. 11, 1999, 17.

94. C. Bedford, *Chem. Ind. (London)*, 1995, 36.

95. R. Baldini, *Chem. Eng. News*, May 8, 1995, 4.

96. E. Kirschner, *Chem. Eng. News*, Apr. 24, 1995, 23.

97. (a) D. Mayo, R.M. Pike, and P.K. Trumper, *Microscale Organic Laboratory*, 3rd ed., Wiley, New York, 1994; (b) M.M. Singh, R.M. Pike, and Z. Szafran, *General Chemistry Micro- and Macroscale Laboratory*, Wiley, New York, 1994; (c) M.M. Singh, K.C. Swallow, R.M. Pike, and Z. Szafran, *J. Chem. Ed.*, 1993, *70*, A39; (d) M.M. Singh, R.M. Pike, and Z. Szafran, Preprints A.C.S. *Div. Environ. Chem.*, 1994, *34*(2), 194; (e) Z. Szafran, R.M. Pike, and J.C. Foster, *Microscale General Chemistry Laboratory with Selected Macroscale Experiments*, Wiley, New York, 1993; (f) Z. Szafran, R.M. Pike, and M.M. Singh, *Microscale Inorganic Chemistry: A Comprehensive Laboratory Experience*, Wiley, New York, 1991; (g) J.L. Skinner, *Microscale Chemistry— Experiments in Miniature*, Royal Society of Chemistry, Cambridge, 1998.

98. M. Freemantle, *Chem. Eng. News*, Feb. 22, 1999, 27.

99. Anon., *Chem. Eng. News*, June 3, 1996.

100. Anon., *Chem. Eng. News*, Mar. 30, 1998, 29.

101. S. Mazrou, H. Kerdjoudj, A.T. Cherif, A. Elmidaoui, and J. Molenat, *New J. Chem.*, 1998, *22*, 355.

102. Anon., *Environ. Sci. Technol.*, 1998, *32*, 119A.

103. (a) N. Chadha, *Chem. Eng. Prog.*, 1994, *90*(11), 32; (b) Office of Technology Assessment, *Industry, Technology and the Environment: Competitive Challenges and Business Opportunities*, Washington, DC, 1994, Chap. 8; (c) K. Martin and T.W. Bastock, eds, *Waste Minimisation: A Chemist's Approach*, Royal Society of Chemistry, Cambridge, 1994; (d) J.H. Siegell, *Chem. Eng.*, *103*(6), 92; (e) M. Venkatesh and C.W. Moores, *Chem. Eng. Prog.*, 1998, *94*(11), 26; (f) S.B. Billatos and N.A. Basaly, *Green Technology and Design for the Environment*, Taylor & Francis, Washington, DC, 1997; (g) M. Venkatesh, *Chem. Eng. Prog.*, 1997, *93*(5), 33; (h) P. Crumpler, *Chem. Eng.*, 1997, *104*(10), 102.

104. (a) K. Fouhy, *Chem. Eng.*, 1995, *102*(1), 41; (b) J. Jarosch, *Chem. Eng.*, 1996, *103*(5), 120; (c) *Chem. Eng.*, 1996, *103*(5), 127; (d) D.M. Carr, *Chem. Eng.*, 1995, *102*(8), 78; (e) B.J. Netzel, *Chem. Eng.*, 1995, *102*(8), 82B; (f) Anon., *Chem. Eng.*, 1995, *102*(11); (g) C. Brown and P. Dixon, *Chem. Eng. Prog.*, 1996, *92*, 42; (h) J.H. Siegell, *Chem. Eng. Prog.*, 1998, *94*(11), 33.

105. E.M. Kirschner, *Chem. Eng. News*, July 10, 1995, 14.

106. Environmental, Health and Safety Progress Report for 1995, Exxon Corp, New York, p. 11.

107. J. Szekely and G. Trapaga, *Technol. Rev.*, 1995, *98*(1), 30.

108. M.A. Martin-Cabrejas, R.M. Esteban, F.J. Lopez-Andreu, K. Waldron, and R.R. Selvendran, *J. Agric. Food Chem.*, 1995, *43*, 662.

109. M. Reisch, *Chem. Eng. News*, Feb. 10, 1997, 9.

110. (a) B. Crittenden and S.T. Kolaczkowski, *Waste Minimization Guide, Inst. Chem. Eng.*, Rugby, UK, 1994; (b) D.F. Ciambrone, *Waste Minimization as a Strategic Weapon*, Lewis Publishers, Boca Raton, FL, 1995; (c) T.E. Higgins, *Pollution Prevention Handbook*, Lewis Publishers, Boca Raton, FL, 1995; (d) D.T. Allen, *Adv. Chem.*

Eng., 1994, *19*, 251; (e) L. Theodore, *Pollution Prevention*, van Nostrand–Reinhold, New York, 1992; (f) L. Theodore, *Pollution Prevention—Problems and Solutions*, Gordon and Breach, Reading, 1994; (g) M.J. Healy, *Pollution Prevention Opportunity Assessments—A Practical Guide*, Wiley, New York, 1998; (h) G.F. Nalven, *Practical Engineering Perspectives—Environmental Management and Pollution Prevention*, American Institute of Chemical Engineers, New York, 1997; (i) D.T. Allen and K.S. Ross, *Pollution Prevention for Chemical Processes*, Wiley, New York, 1997; (j) J.H. Clark, ed., *Chemistry of Waste Minimisation*, Blackie, London, 1995; (k) EPA, *Pollution Prevention Guidance Manual for the Dye Manufacturing Industry*, U.S. Environmental Protection Agency, EPA/741/B-92-001.

111. P.W.N.M. van Leeuwen and C. Claver, eds, *Rhodium-Catalyzed Hydroformylation*, Kluwer Academic Publishers, Dordrecht, 2000.

112. K. Weissermel and H.-J.Arpe, *Industrial Organic Chemistry*, 2nd ed., VCH, Weinheim, 1993, 127–131.

113. B. Cornils and E. Wiebus, *Chemtech*, 1995, *25*(1), 33.

114. B. Fell, C. Schobben, and G. Papadogianakis, *J. Mol. Catal., A: Chem.*, 1995, *101*, 179.

115. (a) SRI International, *Chemical Economics Handbook*, Jan. 1991, under Oxo Chemicals, Menlo Park, CA; (b) H. Bach, R. Gartner, and B. Cornils, *Ullmann's Encyclopedia Industrial Chemistry*, 5th ed., VCH, Weinheim, 1985, *A4*, 450–452.

116. (a) R.A. Sheldon, *Chemtech*, 1994, *24*(3), 38; (b) C.B. Dartt and M.E. Davis, *Ind. Chem. Res.*, 1994, *33*, 2887; (c) J.A. Cusumano, *Chemtech*, 1992, *22*(8), 482; *Appl. Catal. A General*, 1994, *113*, 181; (d) J. Haber, *Pure Appl. Chem.*, 1994, *66*, 1597; (e) A. Mittelman and D. Lin, *Chem. Ind. (London)*, 1995, 694.

117. R.A. Sheldon, *Chemtech*, 1994, *24*(3), 38.

118. (a) N. Bodor, *Chemtech*, 1995, *25*(10), 22; (b) N. Bodor. In: S.C. de Vito, and R.L. Garrett, eds., *Designing Safer Chemicals: Green Chemistry for Pollution Prevention*. A.C.S. Symp. 640, Washington, DC, 1996; (c) see also S.C. de Vito, *Chemtech*, 1996, *26*(11), 34 for more ways to design safer chemicals.

119. L.J. Casarett and J. Doull, eds, *Toxicology—The Basic Science of Poisons*, Macmillan, New York, 1975, 195, 514, 516, 720.

120. National Science and Technology Council, A National R&D Strategy for Toxic Substances and Hazardous and Solid Waste, Sept. 1995, Available from U.S. EPA, Office of Research and Development, Washington, DC, 20460.

121. J.S. Hirschhorn, *Prosperity Without Pollution*, van Nostrand-Reinhold, New York, 1991.

122. J. Boor, Jr., *Ziegler–Natta Catalysts and Polymerization*, Academic Press, New York, 1979.

123. G. Fink, R. Mulhaupt, and H.H. Brintzinger, eds, *Ziegler Catalysis—Recent Scientific Innovations and Technological Improvements*, Springer, Heidelberg, 1995.

124. (a) J. Scheirs and W. Kaminsky, eds, *Metallocene-Based Polyolefins*, Wiley, New York, 2000; (b) T. Takahasha, ed., *Metallocenes in Regio- and Stereoselective Syntheses*, Springer, Heidelberg, 2005; (c) W. Kaminsky, ed., *Macromol. Symp.*, 2006, *236*, 1–258.

125. B. Rieger, L.S. Baugh, S. Kacker, and S. Striegler, eds, *Late Transition Metal Polymerization Catalysis*, Wiley, New York, 2003.

126. T.R. Younkin, E.F. Connor, J.I. Henderson, S.K. Friedrich, R.H. Grubbs, and D.A. Bansleben, *Science*, 2000, *287*, 460.

127. S.-M.Yu, A. Berkefeld, I. Gottker-Schnitmann, G. Muller, and S. Mecking, *Macromolecules*, 2007, *40*, 421.

128. J.S. Hirschhorn, *Chemtech*, 1995, *25*(4), 6.

129. B.M. Rushton, *Chem. News*, Jan. 2, 1995, 2.

2 The Chemistry of Longer Wear

2.1 WHY THINGS WEAR OUT

The first step in achieving longer wear is to have reusable items that can be used many times before they wear out or break and have to be recycled. If the process of wear or breakage is understood, it may be possible to intervene by altering the design or manner of use so that a long service life is obtained.[1]

Metals can rust or corrode. Rubber can harden and crack with age. Plastic may embrittle with age. Colors may fade in the sunlight. These are due to oxidation. Efforts to minimize these processes involve isolation of the object from the environment, by a coating or other means, or by the use of additives that prevent the oxidation. Concrete may deteriorate under the influence of deicing salts. Other objects wear out mechanically. A cutting edge becomes dull. The knees of the pants wear through from abrasion. Stockings snag and trip. Teeth come out of zippers. Plastics may scratch or stain. The rung on the chair may break. Glass and china objects may break if they are dropped or hit. Some plastics are brittle on impact. Biological factors may also be involved. The clothes moth likes to eat woolen garments. Water-based paints, inks, cutting fluids, and such may serve as substrates for the growth of bacteria and fungi. Termites and carpenter ants may eat the wood in houses. Understanding the ecology of these organisms can sometimes provide a means to prevent their entry. Otherwise, biocides are used.

There is often a weak link in an appliance or piece of equipment. A plain steel screw in a stainless steel pot will fail first. The screw is used because stainless steel is harder to machine. A plain steel knife keeps its cutting edge better than one made of stainless steel. A small spring may fatigue and break. A nut may be lost from a bolt. Many nuts are locked on by putting a monomer on the threads, which then polymerizes when no more oxygen is present.[2] This is an anaerobic adhesive. The important thing is for the equipment to be designed for ease of disassembly for easy replacement of the weak link. This will require more screws and fewer rivets, welds, and metal parts encased in plastic. It may require more standardization of sizes of screws or other parts. The second factor is that the spare part should be easy to obtain, preferably locally. A generation ago, it was possible to buy the one screw needed at the neighborhood hardware store. Today, one may have to drive to a mall and buy a prepackaged assortment of two dozen screws. It may be best if the owner can replace the part himself. The next best option would be a local repair service. For more complicated equipment, shipment to a central repair facility might be necessary. Generators for automobiles are handled in this manner. Other weak links include the elastic thread in underwear that fails long before the cloth, and the holes that develop in pockets long before they form anywhere else in the pants. Ideally, the weak link would be redesigned so that it is no longer weak. This may require chemists to devise new and better materials.

Some of the failures are due to the improper choice of materials or poor designs. The cutting edges on saws, planes, chisels, knives, and such will stay sharp longer if they are made of a good tool steel. They should be easy to resharpen. The spring will last longer if it is made of a good tool steel and not overextended in use. It is easy to cast "white metal," but the castings tend to be somewhat brittle and often break with time. A household strainer of stainless steel will last longer than one made of plain steel. Stainless steel knives, forks, and spoons will hold up better than those plated with silver, and there is no tarnish to be taken off. Chutes and trucks carrying rocks will last longer if they are lined with rubber or with ultrahigh molecular weight polyethylene. Levees at New Orleans failed because the force of Hurricane Katrina exceeded that expected by the designers.[3] In nearby Mississippi, houses worth US$400,000 collapsed because they lacked US$20 cross braces. The area has now adopted a building code.

A great many useful objects are made of plastic. The preferred methods of fabrication of many plastics involve melting the plastic and putting it into a mold or forcing it through a spinneret to form a fiber. If this process is done at too high a temperature, the polymer or the additives in it may be degraded. The extra heat is usually applied in an effort to speed up the line. As the molecular weight increases, so does the viscosity and the strength. To obtain the maximum strength, the molecular weight should be high enough to be on a plateau where no further increase in one will affect the other. The usual molding conditions are a compromise between optimal strength and a viscosity that can be handled in the equipment. Adding ultrasound to an extruder can raise the molecular weight above that which can be processed without it. Ultrahigh molecular weight polyethylene (molecular weights of 1 million or more) cannot be processed by these methods. However, when fiber is prepared by a solution method, the fiber is strong enough to be used in bulletproof vests. Further research on cross-linking or chain extension during fabrication may lead to stronger, tougher parts. On the other hand, too much cross-linking can lead to hard, brittle objects. The glass transition temperature is also important. If a rubbery polymer is used below its t_g, it will be rigid instead of rubbery. If a plastic is used above its t_g, it will gradually change its size or shape, a process termed creep.

Additives to the polymers make a world of difference.[4] A brittle polymer, such as polystyrene, can be made to resist impact by the inclusion of little rubber balls a few microns in size. The rubber diffuses and dissipates the force that hits the objects. The trick is to find a way to make the rubber separate in the proper manner during fabrication. The best method involves light grafting to the preformed rubber during polymerization of the other monomers. Preformed rubber particles of the core–shell type have been used to toughen acrylic polymers.[5] Powdered, cross-linked 500 nm rubber particles have been added to polypropylene and ethylene–propylene copolymers to give thermoplastic elastomers.[6] The same principle applies to the macroscopic particles of rubber that make asphalt pavements last longer. Reinforcing fibers and fillers can increase strength. The lifetimes of automobile tires have been increased greatly in the past 30 years by switching to a radial design of the belts and using reinforcing silica fillers in poly(1,3-butadiene) treads.[7] The carcass is often made of a different rubber (e.g., a copolymer of styrene and 1,3-butadiene). Shoe heels are a different story. They are made of highly filled reclaimed rubber. If they were made of the same rubber as the tire tread, wear would not be a problem.

2.2 STABILIZERS FOR POLYMERS

Polymer stability varies greatly. Poly(methyl methacrylate), bisphenol polycarbonate, and poly(vinyl fluoride) last for many years of outdoor exposure. The first two can be used for unbreakable windows in buses, schools, and elsewhere. The problem is that they become scratched. A highly cross-linked scratch-resistant coating is put on automobile headlights of these polymers by radiation curing. Poly(vinyl fluoride) can be used to coat the exteriors of buildings. Clear exterior finishes for wood only last a year or two, much less than pigmented finishes.[8] Polyurethanes made with aliphatic isocyanates are more stable outdoors than those made with aromatic isocyanates. The latter absorbs more ultraviolet light. Polyolefin polymers are subject to oxidation. Polypropylene and poly(1,3-butadiene) could not be used commercially without the addition of stabilizers. A trace of antioxidant is added even before the workup of these polymers when they are made. The radicals from these polymers are stabilized by being tertiary in the first case and allylic in the second case (Schematic 2.1).

Polyethylene, which lacks the many branches, is more stable, but still requires the addition of stabilizers. Polyisobutylene, which is a saturated rubber, is more stable than the unsaturated rubbers. Poly(methyl methacrylate), in which oxidation would have to produce a primary radical, is much more stable. Polyamide degradation begins with the abstraction of a hydrogen from the methylene group next to the nitrogen atom, giving a radical that can then react with oxygen.[9]

A wide variety of stabilizers are used in plastics and rubbers.[10] It is common for more than one to be used in a given polymer. The total amount added, for example, to polypropylene is usually under 1%. Because stabilizers are used up in protecting the polymer, more may have to be added when used items are recycled by remolding.[11] Oxidation of a hydrocarbon is a chain reaction (Schematic 2.2).[12]

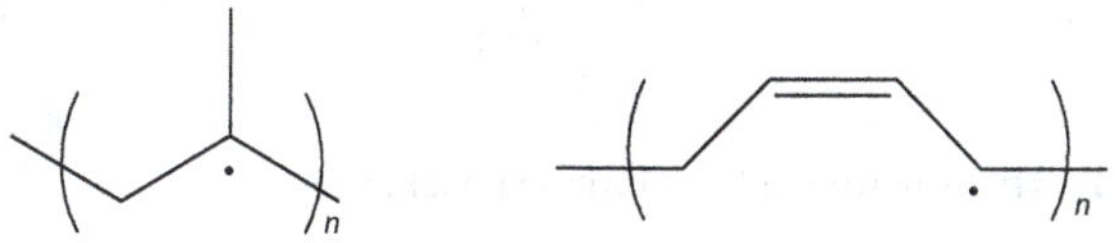

SCHEMATIC 2.1 Polymer radical stabilization

$$R^{\bullet} + O_2 \longrightarrow ROO^{\bullet} \xrightarrow{RH} ROOH + R^{\bullet}$$

$$ROOH + M^{n+} \longrightarrow ROO^{\bullet} + M^{(n-1)+} + H^+$$

$$ROOH + M^{(n-1)+} \longrightarrow RO^{\bullet} + M^{n+} + OH^-$$

$$RO^{\bullet} + RH \longrightarrow ROH + R^{\bullet}$$

SCHEMATIC 2.2 Hydrocarbon oxidation

The initiating radical may be formed by mechanochemical scission during mold-ing[13] or by ozone in the air.[14] If a metal ion that can have multiple valences is present (e.g., a trace of iron from the extruder), the oxidation can be accelerated. In the case of polypropylene, a new radical and, hence, a new hydroperoxide can be formed by backbiting (Schematic 2.3).

An alkoxy radical, formed as previously mentioned, can cleave the chain to form a ketone. The primary radical would promptly abstract a hydrogen atom from something to form a tertiary radical. The extent of degradation of polypropylene is sometimes estimated by measuring the carbonyl content by the infrared spectrum of the sample. If light is present, the ketone can undergo further cleavage.

Antioxidants[15] interrupt these chains by forming radicals that are too weak to abstract hydrogen atoms from carbon atoms. These radical scavengers contain hydrogen atoms that are easily abstracted. The most active are aromatic amines[16] such as those in Schematic 2.4. The radicals formed are stabilized by electron delocalization in a number of resonance structures, some of which are shown in Schematic 2.5. These are staining antioxidants owing to the imine structures that can form. The abstraction of a second hydrogen atom would lead to highly colored quinone imines. The dihydroquinoline cannot do this and is classed as semistaining. Staining antioxidants can be used when the color does not matter, as in gasoline and black rubber tires. Ozone causes cracking in rubber under stress. The cleavage of carbon–carbon double bonds by ozone is well known. Antiozonants tend to be of the aminodiphenylamine type.[17] Waxes that bloom to the surface are used to limit the entry of ozone.

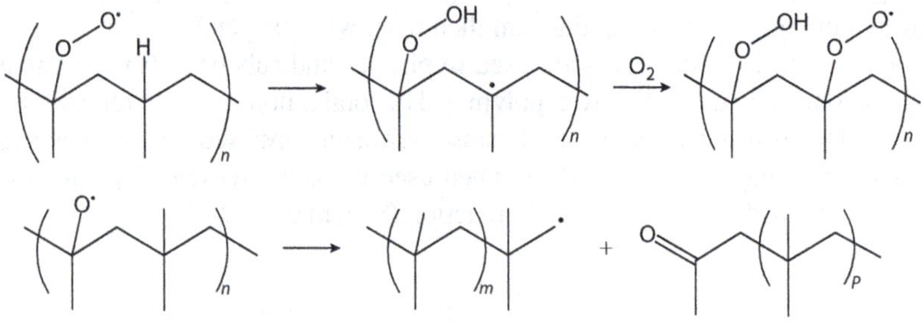

SCHEMATIC 2.3   Hydroperoxide formation via backbiting

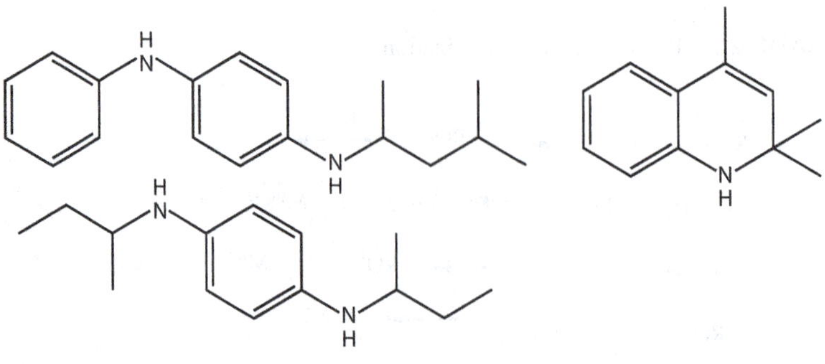

SCHEMATIC 2.4   Aromatic amine radical scavengers

Additional measures have to be taken when light is involved in the degradation of a polymer.[18] Electroluminescent polymers have been encapsulated in resins to keep oxygen and water out and prevent degradation by singlet oxygen.[19] Use of carbon black as a pigment also keeps the light out and prevents photodegradation.[20] Ultraviolet screening agents can also be used to keep the light out. The two types work by photoenolization (Schematic 2.6), which allows the energy to be dissipated as heat without harming the polymer.[21] Typically, R and R′ are long alkyl chains that make the compounds more soluble in polyolefins. Excited states can be quenched by metal chelates, such as those shown in Schematic 2.7 (which has a light green color).

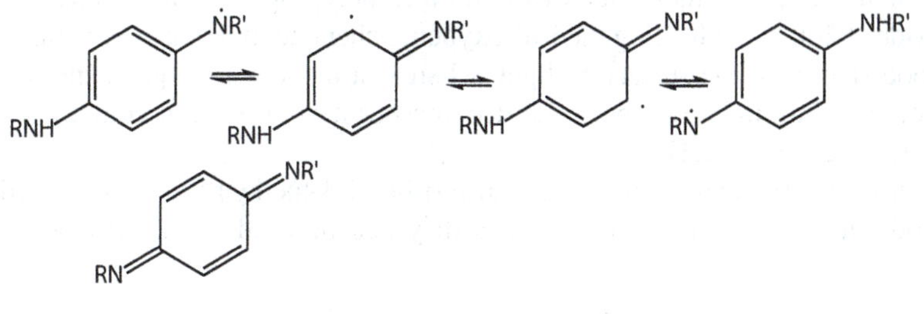

SCHEMATIC 2.5 Radical stabilization via electron delocalization

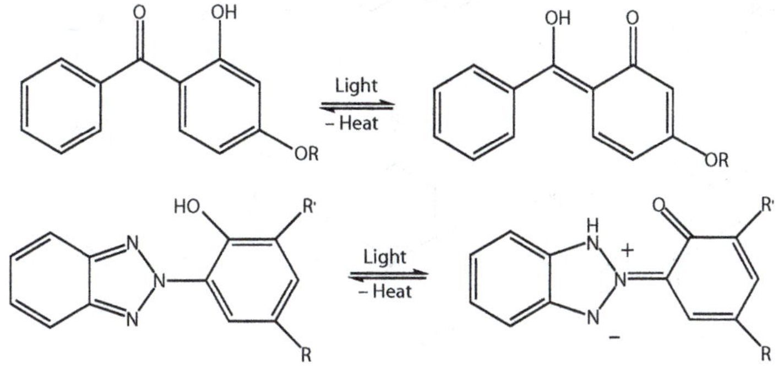

SCHEMATIC 2.6 Photoenolization of ultraviolet screening agents

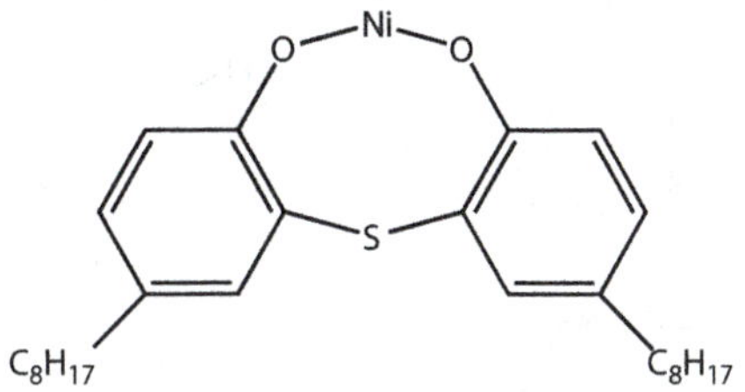

SCHEMATIC 2.7 Metal chelate that quenches an excited state

The best light stabilizers are the hindered amines. A typical one is shown in Schematic 2.8. They are used along with the antioxidants and peroxide decomposers. Their exact mechanism seems to be a matter of debate. One mechanism is shown in Schematic 2.8.[22]

Thus, each molecule destroys more than one radical chain. A variety of analogs of higher molecular weight, often oligomeric polymers, are commercially available for use in fibers and films, where the one shown in Schematic 2.8 might be too volatile.[23] Another method that has been used to reduce volatility is to attach a stabilizer to the polymer that is to be protected.[24] For example, a hindered piperidinol or aminopiperidine was reacted with a graft copolymer of maleic anhydride on polypropylene to form an ester and amide, respectively (Schematic 2.9).[25]

A hindered amine monomer was grafted to polypropylene in the presence of a peroxide at 200°C.[26] Grafting the hydroxybenzophenone monomer (Schematic 2.10) to wood stabilized the surface to light, whereas a monomeric equivalent did not.[27] The hindered amine and a peroxide decomposer have also been combined in one molecule (Schematic 2.11).[28]

Synthetic antioxidants, such as 2.6-di-*tert*-butyl-4-methylphenol, are not allowed in foods in Japan. In the United States they can be used in the wrapper around

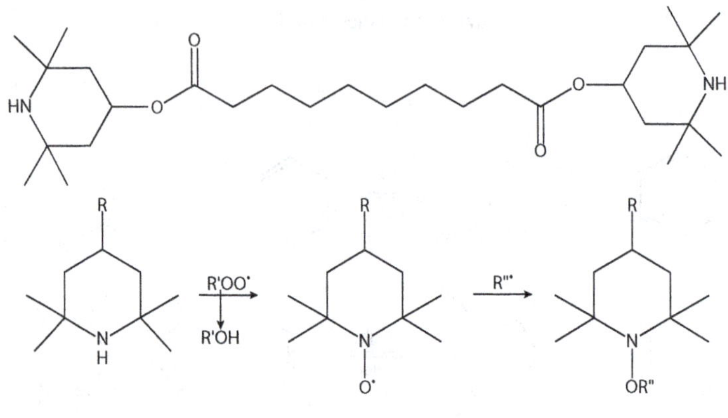

SCHEMATIC 2.8   Hindered amine light stabilizer

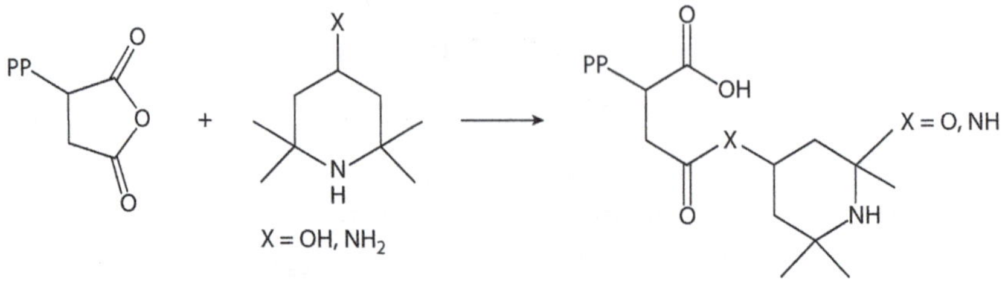

SCHEMATIC 2.9   Reducing the volatility of a polymer

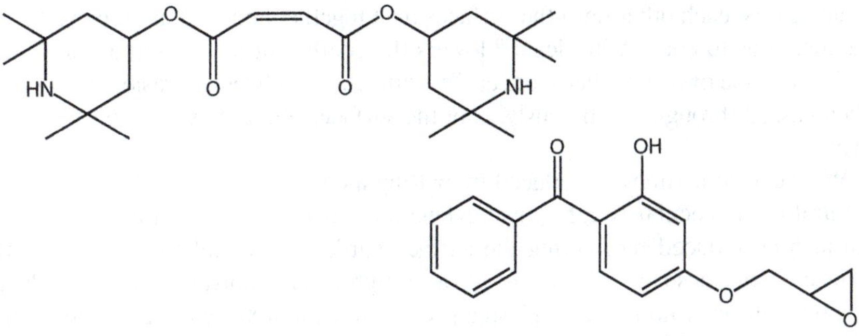

SCHEMATIC 2.10 Monomers used in grafting

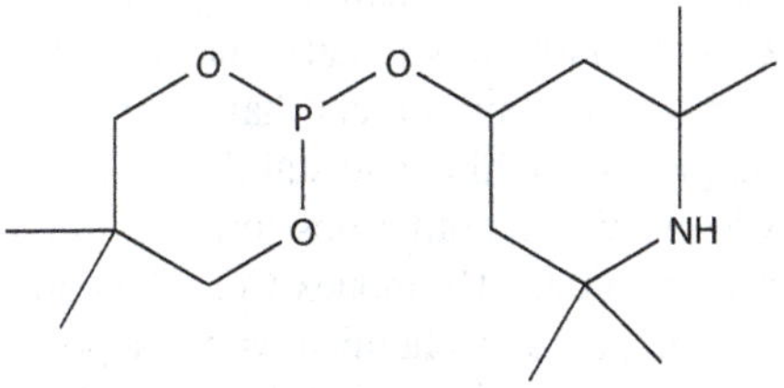

SCHEMATIC 2.11 Hindered amine and peroxide decomposer

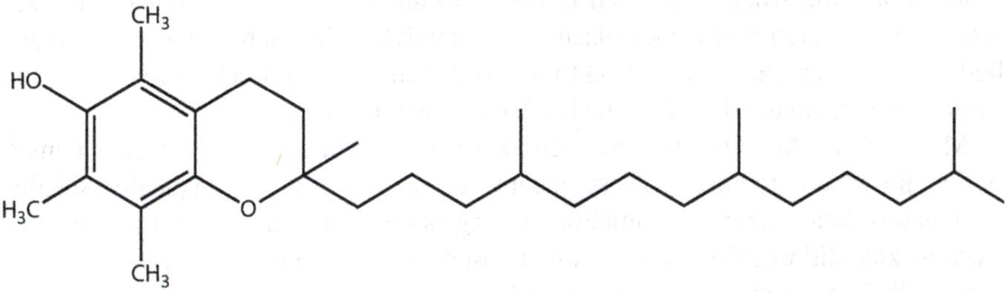

SCHEMATIC 2.12 Vitamin E

the cereal in small amounts, but not in the cereal itself. There is great interest in natural antioxidants that can be used to protect foods.[29] That these natural phenols from plants may scavenge free radicals in the body and thereby reduce the incidence of heart attacks and cancer makes them especially important. Vitamin E (Schematic 2.12) is the one that we must all have in our diet.

2.3 LUBRICATION, WEAR, AND RELATED SUBJECTS

If two objects are rubbed together, they may wear by abrasion. The heat developed in the process may cause them to deteriorate by oxidation or loss of strength, or they may melt. In fact, this is one way to weld two pieces of plastics together. They rotate against each other until the surfaces melt together and then are removed from the apparatus to cool. A lubricant[30] lowers the coefficient of friction, allowing one surface to slide over the other more easily without, or with less, abrasion. It can also, when passed through continuously, cool the surfaces and remove any particles that form.

Wear can sometimes be reduced by putting a coating of polyurethane rubber or of ultrahigh molecular weight polyethylene on a metal surface.[31] Sometimes, wear can also be reduced by making the surface harder. Cross-linked coatings on the surfaces of plastics can do this. The wear of high-density polyethylene was reduced by a thin surface film made by plasma polymerizing a mixture of acetylene, hydrogen, and silane.[32] Harder surfaces were produced on poly(methyl methacrylate) by transamidation with diamines.[33] Another way to do it is to deposit a transparent layer of silica or titania by the sol–gel method. High-energy ion irradiation can be used to make the surfaces of common polymers, such as polyethylene, poly(ethylene terephthalate), polystyrene, nylon, and others, harder than steel.[34] This might provide a way of eliminating the scratching and staining of plastic dinnerware made of melamine–formaldehyde plastics. Another possibility would be to apply a diamond coating by the three-laser method.[35] Diamonex Co. offers amorphous and polycrystalline diamond coatings (prepared by chemical vapor deposition[36]) that increase the wear resistance of plastics, glass, and metals.[37] Diamond-like coatings are hard and slick with a coefficient of friction as low as 0.001 compared with 0.15 for graphite and 0.04 for polytetrafluoroethylene so that lubricants are not needed with them.[38] Plastic dishes coated in this way offer advantages over glass and ceramic dinnerware in not scratching when in use and in not breaking when dropped. Mitsubishi uses acetylene with a radiofrequency plasma to coat poly(ethylene terephthalate) beverage bottles with a diamond-like coating to improve their barrier properties to oxygen and carbon dioxide up to 10-fold.[39] The bottles can be reused.

Most lubricating oils are based on petroleum.[40] For optimum results, numerous additives are required. These include antifoaming agents (typically silicone oils), antioxidants, corrosion inhibitors, detergents or dispersants, metal deactivators (such as zinc dithiocarbamates or dithiophosphates), pour-point depressants [such as a poly(alkyl methacrylate)], viscosity index improvers [such as ethylene–propylene copolymers, hydrogenated diene–styrene copolymers, or a poly(alkyl methacrylate) (from C_4, C_{12}, C_{18} methacrylates)]. Pourpoint depressants lower the temperature at which the oil can be used (e.g., needed in engine oil in parked cars in the winter). They prevent wax from crystallizing out in a form that would interfere with the functioning of the engine. Viscosity index improvers[41] reduce the amount of the decrease in viscosity as the oil heats up. Extreme pressure lubricants, as needed for gears, use phosphorus or sulfur compounds, such as sulfurized polybutene and zinc dialkyldithiophosphates. As one part slides over another, these compounds produce lubricating layers of metal sulfides or phosphides on the metal surfaces.[42] The zinc dialkyldithiophosphate also functions as an inhibitor for corrosion and oxidation. Greases are made by adding 4%–20% of a metal (usually lithium or calcium) fatty acid soap to an oil to thicken it. Inorganic thickeners, such as silica and montmorillonite (a clay), can also be used. A stable suspension of boric acid in motor oil can reduce friction by two thirds and result in a 4%–5% reduction in fuel consumption.[43]

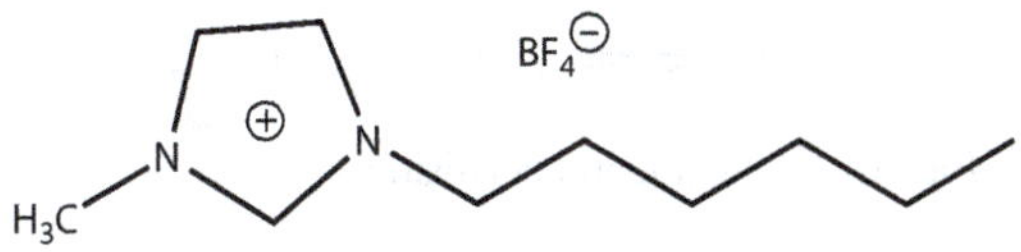

SCHEMATIC 2.13 Ionic liquid lubricant

Synthetic oils[44] are used for more demanding applications, such as use at higher temperatures. Esters of pentaerythritol [$C(CH_2OCOR)_4$, where R averages about C_7] are often used at higher temperatures. Phosphate esters, such as tri(o-tolyl)phosphate, and silicone oils are also used at higher temperatures. Silicone grease is made by dissolving a lithium soap in a silicone oil. Solid lubricants, such as the layered materials graphite and molybdenum disulfide,[45] are used for extreme conditions of temperature or chemical resistance. Hollow nanoparticles of tungsten disulfide outperform these materials and may come into common use.[46] They are made by treating hollow nanoparticles of tungsten(VI) oxide with hydrogen sulfide and hydrogen at 800–850°C.

High-oleic vegetable oils can be used in place of mineral oils.[47] They can also be used as hydraulic fluids. They can be used as food-grade lubricants, are from renewable sources, and are biodegradable, which helps if a spill occurs. Additives are also required for these oils. Ethanolysis of such oils would produce ethyl oleate for use as fuel with the same diesel engine that uses the oils for lubricant and in its hydraulic lines. Oligomeric esters of oleic acid, formed by self-addition of the carboxylic acid groups to the double bond in the presence of a sulfuric acid catalyst, can also be used as biodegradable lubricants.[48] Some ionic liquids are lubricants (Schematic 2.13).[49]

2.4 INHIBITION OF CORROSION

Despite man's best efforts, many objects have their lifetimes shortened by corrosion.[50] One third of chemical plant failures in the United States are due to corrosion.[51] Some notable examples are the Three Mile Island nuclear reactor, a nuclear reactor in Ohio, the Alaska pipeline, and the Motiva (now Valero) oil refinery in Delaware City, DE. The cost of corrosion in the United States is US$300 billion.[52] Corrosion should be monitored to be sure that it is not excessive.[53] Batelle has devised a system with a new layer between the base coat and the top coat that fluoresces under a scanning device if corrosion has started. This can reveal it before it is visible to the naked eye.[54] The corrosion of metals is an electrochemical process (Schematic 2.14).

The forward reaction takes place at the anode, the reverse at the cathode. An electrically conducting pathway is needed to complete the circuit. Corrosion is often localized, forming pits and crevices wherever the passivating layer of oxide is broken by aggressive anions. Galvanic corrosion can occur when two dissimilar metals are joined, one becoming the anode and the other the cathode. Magnesium and zinc have been used as sacrificial anodes to protect steel from rusting in pipelines, ships, hot water heaters, and other equipment. A sprayed zinc anode has been used to protect the steel bars in concrete highways from the effects of deicing salts.[55] Environmental effects such as pH, nature and concentration of anions, the

$$M \rightleftharpoons M^{n+} + ne^-$$

SCHEMATIC 2.14   Electrochemical metal corrosion

concentration of oxygen, and temperature influence the corrosion process. Living organisms, such as bacteria, sometimes produce metabolites that cause corrosion.[56] This is also true of buildings and monuments made of stone, on which biofilms can increase the deposition of pollutants as well as produce metabolic acids.[57] Acid rain is particularly hard on limestone, marble, and sandstone, which is held together by carbonates. Genetically engineered aerobic bacteria form a protective film on surfaces in cooling water systems that cut the corrosion of steel, stainless steel, copper, and brass by 35- to 40-fold, by thwarting the growth of the usual corrosion-causing bacteria.[58] This could be cheaper and less damaging to the environment than treatment with biocides.

The recovery of artifacts from the sunken ship, Titanic, shows the results of corrosion over 89 years at $0°C$ and pH 8.2 under 2.5 miles of water.[59] Aided by bacteria and other agents, steel had corroded to give $FeO(OH)$, Fe_2O_3, and $FeCO_3$. Lead carbonate and lead sulfide were derived from the paint. Brass had corroded to $Cu(OH)Cl$. There was a slow conversion of silver to silver chloride.

A variety of reagents are used to inhibit corrosion.[60] The shift from solvent-based to water-based paints, cleaning solutions, inks, cutting oils, and such will require increased use of such inhibitors. Corrosion can be inhibited in various ways. A reagent may complex with the metal surface and alter the redox potential of the metal or prevent adsorption of aggressive ions. A passivating layer may be formed and stabilized. The diffusion of oxygen to the surface may be inhibited or the oxygen in the surrounding medium may be scavenged. The surface may also be isolated by application of a protective coating of another metal or a polymer.

Chromate ion is an anodic inhibitor for steel that passivates the surface. Because chromate is toxic, it is being replaced[61] with molybdates. Chromate works in the absence of oxygen. The replacements, molybdate and tungstate, require oxygen to be effective. Cerium and magnesium[62] compounds are alternatives for chromate for the corrosion protection of aluminum. Alkylphosphonic acids perform as well as chromate and have superior adhesion.[63] Silanes, such as 1,2-ethylenebis(triethoxysilane), are also being studied as replacements for chromate.[64] A trimethylsilane plasma has been used to put a base coat on aluminum with no need for chromium.[65] Layer-by-layer deposition of MCM 22/silica composite films gave corrosion resistance to aluminum comparable to chromate.[66] Corrosion-inhibiting pigments may contain chromate or lead.[67] Red lead, Pb_3O_4, was a popular pigment in corrosion-inhibiting primers for steel. These are being replaced by calcium and zinc molybdates or phosphomolybdates. Selective precipitation of calcium or magnesium carbonates on cathodic areas or formation of phosphate films restricts access of oxygen to the metal surface. Hydrazine and sodium sulfite are used to scavenge the oxygen from boiler feedwater.[68]

Organic inhibitors may absorb on the whole surface. In hydrophobic fatty amines, the amine coordinates with the metal and the long carbon chains stick out from the surface. Carboxymethylated fatty amines (i.e., *N*-substituted glycines)

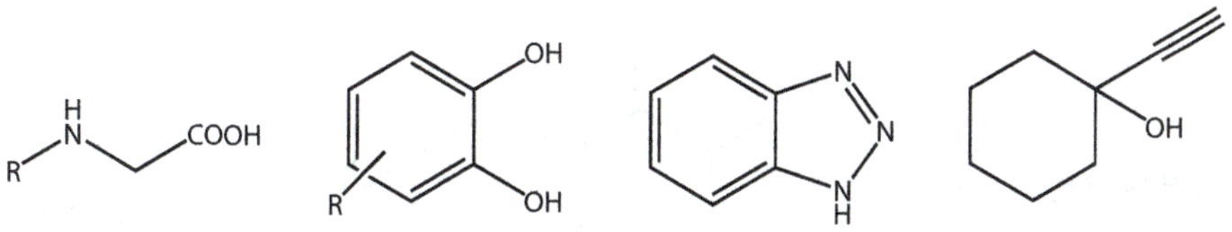

SCHEMATIC 2.15 Corrosion inhibitors

chelate the metal on the surface. If a chelating agent forms an insoluble surface coating, it may be a good corrosion inhibitor. If it forms a soluble chelate, corrosion may be enhanced. Alkylcatechols also inhibit the corrosion of steel. [Catechol forms strong chelates with Fe(III).] Benzotriazole protects copper from corrosion by formation of a copper(I) chelate. Ethynylcyclohexanol absorbs to steel and then polymerizes (Schematic 2.15).

2.5 MENDING

Mending and patching can make many items last longer so that they do not need to be replaced as soon. The chemistry may have to be different when this has to be done at room temperature on something that was done at elevated temperature in the factory. A scratch or chip in the baked-on finish on a car or an appliance may have to be touched up with a different coating dissolved in a solvent. Epoxy body solder for auto body repairs has to cure at room temperature. The challenge, of course, is to make the repair as strong and durable as the original material, and to do so even if the surface preparation has been less than perfect. Plastic wood and patching plaster are familiar to many persons. Self-healing polymers have been devised.[69] An epoxy resin contained a ruthenium ring-opening catalyst and microcapsules of dicyclopentadiene, which polymerized when broken by a crack.[70] Another contained a microencapsulated mercaptan and a microencapsulated glycidyl ester (Schematic 2.16).[71]

The iron-on patch was a big step forward in repairing clothing. The hot-melt adhesive used on it must adhere well to the fabric initially and through repeated laundering or dry cleaning, must not soften in the clothes drier, must not be damaged by repeated flexing in use, and must be applicable with a home iron under conditions that may not be well-controlled. One material that is used is a terpolyamide composed of 6, 6/36, and 6/12 nylons that soften at 100–150°C.[72] (The numbers refer to the number of carbon atoms in the ingredients, with those in the diamine being given first.) Such hot-melt adhesives are also available as fusible webs that can be used to reinforce fabrics.[73] Although these adhesives are not altered chemically when they are applied, it should be possible to develop some that will cure to tougher polymers on application. A mixture of microencapsulated diepoxide and curing agent would cure when the capsules were ruptured by heat or pressure. Some adhesives, such as that sold for installing Velcro fasteners, cure by evaporation of solvent. Velcro fasteners on stickyback tape (that contains no solvent) are now available commercially. "Shoe Goo II" applied to worn spots on the shoes can extend the life of shoes appreciably. It appears to be a thick solution of a rubber in a mixture of propyl acetate and petroleum distillate. Similar solutions of polymers are available for mending broken

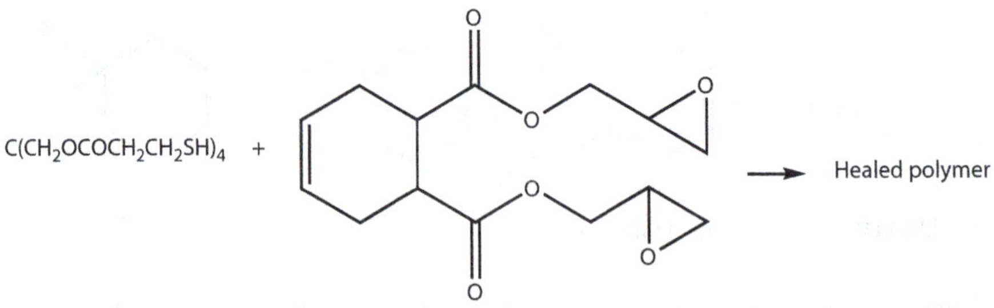

SCHEMATIC 2.16 Self-healing polymer formation

china and other items. Adhesive tapes on release paper can be removed and used to hem garments (i.e., to replace sewing). Glue sticks are also available for trimming, and putting in hems, zippers, and what not. Some of these techniques should be applicable to putting replacement liners in pant pockets.

The weak links in clothing include the knees, elbows, pockets, and zippers. Ideally, these should be made more durable in the first place. Zippers with plastic (properly stabilized) teeth large enough so as not to catch on the adjacent fabric, on a strong fabric backing, should not wear out before the rest of the garment. Stronger and thicker, if necessary, cloth could be used for the pockets. Areas of wear such as knees, elbows, and seats could be reinforced with strong fabric. When the sharp points of shirt collars are badly worn, the collars can be removed and the shirts worn without them. Clothing that wears out beyond the ability to patch it can sometimes be used for other purposes.[74] Long pants with worn-out knees can be converted to shorts. A long-sleeved sweater with worn-out elbows can be converted to a short-sleeved one. A dress can be cut off to a blouse. Worn-out sweaters can be converted to mittens. Portions of one worn-out garment that are still good can be used to patch another. Sections of adult clothing that are still good can be converted to clothing for children. In earlier years, scraps of clothing went into patchwork quilts or braided rugs. Old rags are also useful for cleaning around the house.

2.6 THE FUTURE

Chemistry can provide stronger, more durable materials that will last longer. Studies of the ways in which polymers and other materials degrade and wear out may lead to new ways of intercepting the processes. This may mean the design of new stabilizer systems, new methods of construction, or new polymers.

The problem of overconsumption by the materials-oriented societies of industrialized nations is a difficult one that must be solved for a sustainable future.[75] A possible approach is to make the real cost of an item apparent to the purchaser at the time of purchase. This might encourage the purchaser to buy the item based on the lowest cost per unit of performance instead of the lowest initial cost. The label might contain items such as cost of the container, cost of the product, cost for each use, expected lifetime, cost per month or year, amount of use to be expected before resharpening is needed, and more. This would show that the cost of an item packaged in a throwaway container is higher than that of one packaged in a reusable container. A customer would have a choice of paying 10 cents for a throwaway glass bottle or 1 cent per use for a refillable glass bottle. A saw of poor steel that had to be sharpened every few months would be less desirable than one of good steel that lasted 10 years before resharpening was needed. A label-rating system might also be used. This might be a list of features of good construction with those in the item for purchase being checked. The presence of dovetailed joints on drawers in furniture or double seams on clothing are examples.

The jobs that will be lost in the manufacture of throwaway items will be counterbalanced by an increased number of jobs in maintenance and repair. Many items are discarded today because it is cheaper to buy a new one than to repair the old one. This calls for research on new automated methods of repair as well as on new designs that make repair simple. Increasing the cost of disposal may also help.

PROBLEMS

2.1   I Lost My Pants

Elastic waists are common in clothing, especially in underwear, pajamas, and socks. The elastic is made from "cut thread," which is manufactured by cutting narrow strips from a sheet of natural rubber, which are then wrapped with textile fibers. The problem is that the elastic wears out long before the fabric of the garment. Is this a case of built-in obsolescence designed by the manufacturer to increase sales? What can a poor hungry student do when this problem arises? Is there any way that he or she might mend or fix the clothing? Should they go to a thrift shop and purchase items with flamboyant colors as replacements?

Suppose that you are a rubber or textile chemist. How might you redesign the rubber or garment, or both, to eliminate the issue? Are there alternative ways to hold up your pants? If you are successful, your company can launch an advertising campaign for garments that last several times as long at no extra cost. This should corner the market share, especially if the improvements are patented.

2.2   My Pantyhose Ran Faster than the Runners in the Boston Marathon

Americans produce more waste per capita than the people of any other nation, largely due to an addiction to the single-use, throwaway habit. Should pantyhose fall into this category, or should it be more sustainable and worn repeatedly? They often run after the first or second wearing. Propose a solution to this wasteful problem.

2.3   The Clothes Horse

The average person buys 64 garments and seven pairs of shoes each year. Americans throw away an average of ten pounds of clothing each year. Of that, only 15% is recovered or recycled. Only 20% of the clothes that are donated to charity thrift shops actually get sold as there is just too much to sell. About half of second-hand clothing gets turned into fibers and wiping rags. The rest is shipped abroad for use there. This means that you may see an Arizona State University T-shirt in Costa Rica. Why do Americans do this? Is it a change in size or shape, or is it just fashion? Do people just want something new or different? How might clothes be modified to become more adaptable and/or last longer?

2.4 Tires on and in Rubber

A 1994 law mandated the use of ground rubber tires in asphalt. This makes the asphalt last longer, resist cracking, require less maintenance and reduces road noise, so that it is cheaper in the long run. It can save up to US$50,000 per lane mile. It is now the largest single market for scrap tires at about 12 million tires per year. California and Arizona use the most with some also being used in Florida. Why has the law not been enforced elsewhere, and why is the use in warmer states? What problems may there be with the concept?

2.5 An Opinion Poll on Clothing and Other Items

Consider the following items and state your preference for ones that are "flat" or "shiny." People and birds often prefer "shiny." Birds have been known to chew up shiny aluminum labels on plants.

1. (a) Automobile body, lawn mower, refrigerator, clothes washer; (b) automobile bumper, furniture; (c) floor; (d) finger nails; (e) shoes; (f) lighting fixtures; (g) earrings; (h) seats of pants.
2. Should pants have creases in them?
3. Could you get used to the rumpled look in clothing?
4. Should businessmen and women dress in suits?
5. For exterior siding on houses, do you prefer plain brick, painted brick, natural wood, painted wood, aluminum, or poly(vinyl chloride)?
6. Do you own much clothing that requires dry cleaning?
7. How should grass clippings be bagged?
8. Would you use a hair spray containing an organic solvent?
9. Do you use a solar clothes dryer? Should they be allowed in the suburbs?
10. Do you own any of the following: leaf blower, electric hedge trimmer, chainsaw, camper vehicle, powered lawn edger?

REFERENCES

1. (a) M.E. Eberhart, *Why Things Break*, Harmony Books, New York, 2003; (b) D.J. Wulpi, *Understanding How Components Fail*, 2nd ed., ASM International, Materials Park, OH, 1999; (c) A.F. Liu, *Mechanics and Mechanisms of Fracture: An Introduction*, ASM International, Materials Park, OH, 2005.
2. S.C. Temin, *Encyclopedia of Polymer Science and Engineering*, 2nd ed., Wiley, New York, 1985, *1*, 570.
3. E. Kintisch, *Science*, 2005, *310*, 953.
4. (a) M.P. Stevens, *J. Chem. Educ.*, 1993, *70*, 445, 535, 713; (b) R. Gachter and H. Muller, *Plastics Additives Handbook*, 4th ed., Hanser–Gardner, Cincinnati, OH, 1993; (c) S. Al-Malaiba, A. Golovoy, and C.A. Wilbie, eds, *Chemistry and Technology of Polymer Additives*, Blackwell Science, Oxford, 1999.
5. P.A. Lovell, *Trends Polym. Sci.*, 1996, *4*, 264.
6. X. Zhang, G. Wei, Y. Liu, J. Gao, Y. Zhu, Z. Song, F. Huang, M. Zhang, and J. Qiao, *Macromol. Symp.*, 2003, *193*, 261.

7. (a) U. Goerl, A. Hunsche, A. Mueller, and H.G. Koban, *Rubber Chem. Technol.*, 1997, *70*, 608; (b) A.S. Hashim, B. Azahari, Y. Ikeda, and S. Kohjiya, *Rubber Chem. Technol.*, 1998, *71*, 289; (c) J.T. Byers, *Rubber World*, 1998, *218*(6), 38, 40, 44, 46.

8. T. Daniel, M.S. Hirsch, K. McClelland, A.S. Ross, and R.S. Williams, *JCT-Coatings Technol.*, 2004, *1*(9), 42.

9. (a) B. Lanska, *Angew. Makromol. Chem.*, 1997, *252*, 139; (b) P.N. Thanki and R.P. Singh, *J. Macromol. Sci. Rev. Macromol. Chem.*, 1998, *C38*, 595.

10. (a) M.P. Stevens, *J. Chem. Educ.*, 1993, *70*, 535; (b) M. Dexter, *Kirk–Othmer Encyclopedia of Chemical Technology*, 4th ed., Wiley, New York, 1992, *3*, 424; (c) R.L. Clough, N.C. Billingham, and K.T. Gillen, eds, *Polymer Durability: Degradation, Stabilization and Lifetime Prediction*, ACS Adv. Chem. Ser. 249, Washington, DC, 1996; (d) G.E. Zaikov, ed., *Degradation and Stabilization of Polymers – Theory and Practice*, Nova Science, Commack, New York, 1995; (e) S.H. Hamid, M.B. Amin, and A.G. Maadhah, eds, *Handbook of Polymer Degradation and Stabilization*, Dekker, New York, 1992; (f) N.S. Allen and M. Edge, *Fundamentals of Polymer Degradation and Stabilization*, Elsevier, London, 1992; (g) G. Scott, *Mechanism of Polymer Degradation and Stabilization*, Elsevier, London, 1990; *Atmospheric Oxidation and Antioxidants*, Elsevier, Amsterdam, 1993; (h) J. Pospisil and S. Nespurek, *Polym. Degrad. Stabil.*, 1995, *49*, 99; *Macromol. Symp.*, 1997, *115*, 143; (i) H. Zweifel, *Macromol. Symp.*, 1997, *115*, 181; (j) W.D. Habicher, I. Bauer, K. Scheim, C. Rautenberg, A. Lossack, and K. Yamaguchi, *Macromol. Symp.*, 1997, *115*, 93; (k) J.A. Kuczkowski, *Rubber World*, 1995, *212*(5), 19; (l) G. Scott, *Antioxidants in Science, Technology, Medicine and Nutrition*, Albion, Chichester, 1997; (m) G. Pritchard, ed., *Plastics Additives: An A-Z Reference*, Chapman & Hall, New York, 1998; (n) M. Bolgar, J. Hubball, S. Meronek, and J. Groeger, *Handbook for Chemical Analysis of Plastic and Polymer Additives*, CRC Press, Boca Raton, FL, 2008.

11. (a) R. Pfaendner, H. Herbst, K. Hoffmann, and F. Sitek, *Angew. Makromol. Chem.*, 1995, *232*, 193; (b) J. Pospisil, F.A. Sitek, and R. Pfaendner, *Polym. Degrad. Stabil.*, 1995, *48*, 351.

12. E.T. Denisov and V.V. Azatyan, *Inhibition of Chain Reactions*, Gordon and Breach, London, 2000.

13. G. Scott, *Polym. Degrad. Stabil.*, 1995, *48*, 315.

14. F. Gugumus, *Polym. Degrad. Stab.*, 1998, *62*, 403.

15. (a) Y.A. Shlyapnikov, S.G. Kiryushkin, and A.P. Mar'in, *Antioxidative Stability of Polymers*, Taylor & Francis, London, 1996; (b) D.W. Hairston, *Chem. Eng.*, 1996, *103*(5), 71; (c) E.T. Denisov and I.B. Afanasev, *Oxidation and Antioxidants in Organic Chemistry and Biochemistry*, CRC Press, Boca Raton, FL, 2005.

16. (a) B. Lanska, *Polym. Degrad. Stabil.*, 1996, *53*, 89, 99; (b) J. Pospisil, *Adv. Polym. Sci.*, 1995, *124*, 87.

17. R.P. Lattimer, C.K. Rhee, and R.W. Layer, *Kirk–Othmer Encyclopedia of Chemical Technology*, 4th ed., Wiley, New York, 1992, *3*, 448.

18. (a) J.F. Rabek, *Photodegradation of Polymer*, Springer, Heidelberg, 1996; (b) J.F. Rabek, *Polymer Photodegradation—Mechanisms and Experimental Methods*, Chapman & Hall, London, 1995; (c) R.L. Clough and S.W. Shalaby, eds, *Irradiation of Polymers: Fundamentals and Technological Applications*, ACS Symp. 620, Washington, DC, 1996; (d) V.Y. Shlyapintokh, *Photochemical Conversion and Stabilization of Polymers*, Hanser–Gardner, Cincinnati, OH, 1985; (e) A. Faucitano, A. Buttafava, G. Camino, and L. Greci, *Trends Polym. Sci.*, 1996, *4*(3), 92; (f) F. Gugumus, *Polym. Degrad. Stabil.*, 1994, *44*, 273, 299; 1995, *50*, 101; (g) H.B. Olayan, H.S. Hamid, and E.D. Owen, *J. Macromol. Sci. Rev. Macromol. Chem.*, 1996, *C36*, 671; (h) A. Tidjani, *J. App. Polym. Sci.*, 1997, *64*, 2497; (i) C. Decker, Biry, and K. Zahouily, *Polym. Degrad. Stabil.*, 1995,

49, 111; (j) K. Kikkawa, *Polym. Degrad. Stabil.*, 1995, *49*, 135; (k) A. Rivaton, *Polym. Degrad. Stabil.*, 1995, *49*, 163; (l) M.C. Celina and R.A. Assink, *Polymer Durability and Radiation Effects*, ACS Symp., 978, Oxford University Press, New York, 2007.

19. (a) B.H. Cumpston and K.F. Jensen, *Trends Polym. Sci.*, 1996, *4*(5), 151; (b) L. Peeters and H.J. Giese, *Trends Polym. Sci.*, 1997, *5*(5), 161.

20. (a) J. Mwila, M. Miraftab, and A.R. Horrocks, *Polym. Degrad. Stabil.*, 1994, *44*, 351; (b) A.R. Horrocks and M. Liu, *Macromol. Symp.*, 2003, *202*, 199.

21. M. Zayat, P. Garcia-Parejo, and D. Levy, *Chem. Soc. Rev.*, 2007, *36*, 1270.

22. J.-J. Lin, M. Cuscurida, and H.G. Waddill, *Ind. Eng. Chem. Res.*, 1997, *36*, 1944.

23. F. Gugumus, *Polym. Degrad. Stabil.*, 1998. *60*, 99, 119; 1999, *66*, 133.

24. J. Pospisil, W.D. Habicher, and S. Nespurek, *Macromol. Symp.*, 2001, *164*, 389.

25. R. Mani, N. Sarkar, R.P. Singh, and S. Sivaram, *J. Macromol. Sci. Pure Appl. Chem.*, 1996, *A33*, 1217.

26. J. Malik, G. Ligner, and L. Avar, *Polym. Degrad. Stabil.*, 1998, *60*, 205–213.

27. M. Kiguchi and P.D. Evans, *Polym. Degrad. Stabil.*, 1998, *61*, 33.

28. K. Schweltlick and W.D. Habicher, *Angew. Makromol. Chem.*, 1995, *232*, 239.

29. (a) J.T. Kumpulainen and J.T. Salonen, *Natural Antioxidants and Food Quality in Atherosclerosis and Cancer Prevention*, Royal Society of Chemistry, Cambridge, 1997; (b) F. Shahidi, ed., *Natural Antioxidants: Chemistry, Health Effects and Applications*, American Oil Chemists Society, Champaign, IL, 1996; (c) G. Scott, *Chem. Br.*, 1995, 31(11), 879; (d) I. Kubo, *Chemtech*, 1999, 29(8), 37; (e) E. Haslam, *Practical Polyphenolics: From Structure to Molecular Recognition and Physiological Action*, Cambridge University Press, Cambridge, 1998.

30. (a) E.R. Booser, ed., *Handbook of Lubrication and Tribology*, CRC Press, Boca Raton, FL, 1994; (b) E. Rabinowicz, *Friction and Wear of Materials*, 2nd ed., Wiley, New York, 1995; (c) K.C. Ludema, Friction, *Wear and Lubrication–A Textbook in Tribology*, CRC Press, Boca Raton, FL, 1996; (d) *ASM Handbook 18, Friction, Lubrication and Wear Technology*, ASM International, Materials Park, OH, 1992; (e) D. Klamann, *Ullmann's Encyclopedia Industrial Chemistry*, 5th ed., VCH, Weinheim, 1990, *A15*, 423–518; (f) E.R. Booser, *Kirk–Othmer Encyclopedia of Chemical Technology*, 4th ed., Wiley, New York, 1995, *15*, 463; (g) R. Glyde, *Chem. Br.*, 1997, *37*(7), 39; (h) T. Mang and W. Dresell, eds, *Lubricants and Lubrication*, Wiley-VCH, Weinheim, 2001; (i) A. Shanley, *Chem. Eng.*, 2000, *107*(7), 33; (j) A. Fisher and K. Robzin, *Wear and Wear Protection*, Wiley-VCH, Weinheim, 2009.

31. R. Chattopadhyay, *Surface Wear: Analysis, Treatment and Prevention*, ASM International, Materials Park, OH, 2001.

32. X. Zou, X. Yi, and Z. Fang, *J. Appl. Polym. Sci.*, 1998, *70*, 1561.

33. J. Pavlinec, M. Lazar, and I. Janigova, *J. Macromol. Sci. Pure Appl. Chem.*, 1997, *A34*, 81.

34. (a) R. Dagani, *Chem. Eng. News*, Jan. 9, 1995, 24; (b) E.H. Lee, L.G.R. Rao, and L.K. Mansur, *Trends Polym. Sci.*, 1996, *4*, 229.

35. (a) R. Dagani, *Chem. Eng. News*, Dec. 15, 1997, 11; (b) R.F. Service, *Science*, 1997, *278*, 2057.

36. M.N.R. Ashfold and P.W. May, *Chem. Ind. (Lond.)*, 1997, 505.

37. (a) Anon., *R&D (Cahners)*, 1998, *40*(10), 180; (b) Diamonex Performance Products, Allentown, PA, August 1998.

38. Anon., *Chem. Eng. News*, Jan. 18, 2007, 48.

39. G. Ondrey, *Chem. Eng.*, 2004, *111*(2), 16.

40. (a) S. Ritter, *Chem. Eng. News*, Mar. 13, 2006, 38; (b) A. Shanley, *Chem. Eng.*, 2000, *107*(7), 33.

41. M.K. Mishra and R.G. Saxton, *Chemtech*, 1995, *25*(4), 35.

42. G. Smith, *Chem. Br.*, 2000, *36*(4), 38.

43. M. Whitfield, *Chem. Ind. (Lond.)*, Aug. 27, 2007, 10.
44. (a) P.M. Morse, *Chem. Eng. News*, Sept. 7, 1998, 21; (b) L. Rudnick and R. Shubkin, *Synthetic Lubricants and High-Performance Functional Fluids*, Marcel Dekker, New York, 1999.
45. Anon., *Chem. Eng. News*, Sept. 18, 2000, 47.
46. (a) R. Dagani, *Chem. Eng. News*, June 23, 1997, 10; (b) L. Rapoport, S.Y. Bilik, Y. Feldman, M. Momyonfer, S.R. Cohen, and R. Tenne, *Nature*, 1997, *387*, 791.
47. (a) K. Lal, *Preprints ACS Div. Environ. Chem.*, 1994, *34*(2), 474; (b) Anon., *R&D (Cahners)*, 1996, *38*(13), 9; (c) D. Hariston, *Chem. Eng.*, 1994, *101*(8), 65; (d) G. Parkinson, *Chem. Eng.*, 1996, *103*(11), 21; (e) S.Z. Erhan and J.M. Perez, *Biobased ndustrial Fluids and Lubricants*, Am. Oil Chem. Soc., Champaign, IL, 2002; (f) S. Boyde, *Green Chem.*, 2002, *4*, 293–307.
48. (a) T.A. Isbell, H.B. Frykman, T.P. Abbott, J.E. Lohr, and J.C. Drozd, *J. Am. Oil Chem. Soc.*, 1997, *74*, 473; (b) C. Cermak, *Agric. Res.* 2007, *55*(4), 23.
49. (a) C. Ye, W. Liu, and L. Yu, *Chem. Commun.*, 2001, 2244; (b) F. Zhou, Y. Liang, and W. Liu, *Chem. Soc. Rev.*, 209, 38, 2590.
50. (a) P.J. Moran and P.M. Natishan, *Kirk–Othmer Encyclopedia of Chemical Technology*, 4th ed., Wiley, New York, 1993, *7*, 548; (b) D.A. Jones, *Principles and Prevention of Corrosion*, Macmillan, New York, 1992; (c) F. Mansfeld, ed., *Corrosion Mechanisms*, Dekker, New York, 1987; (d) P. Marcus and J. Oudar, *Corrosion Mechanisms in Theory and Practice*, Dekker, New York, 1995; (e) *ASM Handbook 13, Corrosion*, ASM International, Materials Park, OH, 1987; (f) D. Talbot and J. Talbot, *Corrosion Science and Technology*, CRC Press, Boca Raton, FL, 1997; (g) P.E. Elliott, *Chem. Eng. Prog.*, 1998, *94*(5), 33; (h) T.W. Swaddle, *Inorganic Chemistry—An Industrial and Environmental Perspective*, Academic, San Diego, CA, 1996; (i) T.E. Graedel and C. Leygraf, *Atmospheric Corrosion*, Wiley, New York, 2000; (j) Z. Ahmad, *Principles of Corrosion Engineering and Corrosion Control*, Butterworth-Heineman, Burlington, MA, 2006; (k) J.R. Davis, *Corrosion: Understanding the Basics*, ASM International, Materials Park, OH, 2000; (l) A.S. Khanna, *Introduction to High Temperature Oxidation and Corrosion*, ASM International, Materials Park, OH, 2002; (m) C. Shargay and C. Spurrell, *Chem. Eng.*, 2003, *110*(5), 42; (n) S.A. Bradford, *Corrosion Control*, 2nd ed., ASM International, Materials Park, OH, 2001; (o) A.H. Tullo, *Chem. Eng. News*, Sept. 17, 2007, 20; (p) R.W. Revie, *Corrosion and Corrosion Control*, 4th ed., Wiley, Hoboken, NJ, 2008.
51. C. Punckt, M. Bolscher, H.H. Rotermund, A.S. Mikhailov, L. Organ, N. Budiansky, J.R. Scully, and J.L. Hudson, *Science*, 2004, *305*, 1133.
52. R. Marshall, *Chem. Eng.*, 2008, *115*(8), 5.
53. R.D. Kane, *Chem. Eng.*, 2007, *114*(6), 34.
54. (a) Anon., *Chem. Eng. News*, Feb. 2, 2009, 16; (b) Anon., *Chem. Eng. News, Oct. 26, 2009, 24.
55. (a) J.E. Bennett, *Chem. Eng. Prog.*, 1998, *94*(7), 77; (b) T.H. Lewis, Jr., *Chem. Eng. Prog.*, 1999, *95*(6), 55.
56. (a) Anon., *Chem. Eng. News*, Mar. 17, 1997, 29; (b) B.J. Little, R.I. Ray, and P.A. Wagner, *Chem. Eng. Prog.*, 1998, *94*(9), 51; (c) J. Jass and H.M. Lappin–Scott, *Chem. Ind. (Lond.)*, 1997, 682; (d) S.J. Yuan, S.O. Pehkonen, Y.P. Ting, E.T. Kang, and K.G. Neoh, *Ind. Eng. Chem. Res.*, 2008, *47*, 3008.
57. (a) P. Young, *Environ. Sci. Technol.*, 1996, *30*, 206A; *New Sci.*, 1996, *152*(2054), 36; (b) G. Allen and J. Beavis, *Chem. Br.*, 1996, *32*(9), 24.
58. G. Parkinson, *Chem. Eng.*, 1998, *105*(11), 21.
59. (a) M. Freemantle, *Chem. Eng. News*, Oct. 17, 1994, 49; (b) P. Stoffyn–Egli and D.E. Buckley, *Chem. Br.*, 1995, *31*(7), 551.

60. (a) B.G. Clubley, ed, *Chemical Inhibitors for Corrosion Control*, R. Soc. Chem. Spec. Publ. 71, Cambridge, 1990; (b) D.W. Hairston, *Chem. Eng.*, 1996, *103*(3), 65; (c) Y.I. Kuznetsov, *Organic Inhibitors of Corrosion of Metals*, Plenum, New York, 1996.
61. (a) R.L. Twite and G.P. Bierwagen, *Prog. Org. Coat.*, 1998, *33*(2), 91; (b) A. Barbucci, M. Delucchi, and G. Cerisola, *Prog. Org. Coat.*, 1998, *33*(2), 131.
62. G. Ondrey, *Chem. Eng.*, 2006, *113*(12), 13.
63. (a) I. Maege, E. Jaehne, A. Henke, H.-J.P. Adler, C. Braun, C. Jung, and M. Stratmann, *Macromol. Symp.*, 1997, *126*, 7; (b) K.S. Betts, *Environ. Sci. Technol.*, 1999, *33*, 87A.
64. W.J. van Ooij and T. Child, *Chemtech*, 1998, *28*(2), 26.
65. H.K. Yasuda, Q.S. Yu, C.M. Reddy, C.E. Moffitt, and D.M. Wieliczka, *J. Appl. Polym. Sci.*, 2002, *85*, 1387, 1443.
66. J. Choi, Z. Lai, S. Ghosh, D.E. Beving, Y. Yan, and M. Tsapatsis, *Ind. Eng. Chem. Res.*, 2007, *46*, 7096.
67. C. Simpson, *Chemtech*, 1997, *27*(4), 40.
68. B. Buecker, *Chem. Eng.*, 2008, *115*(2), 30.
69. (a) D.Y. Wu, S. Meure, and D. Solomon, *Prog. Polym. Sci.*, 2008, *33*(5), 479; (b) B. Ghosh and M.W. Urban, *Science*, 2009, 323, 1458; (c) Anon., *Chem. Eng. News*, June 1, 2009, 14.
70. S. Stinson, *Chem. Eng. News*, Feb. 19, 2001, 13.
71. Y.C. Yuan, M.Z. Rong, M.Q. Zhang, J. Chen, G.C. Yang, and X.M. Li, *Macromolecules*, 2008, *41*, 5197.
72. W. Haller, G. Tauber, W. Dierichs, J. Herold, and W. Brockmann, *Ullmann's Encyclopedia of Industrial Chemistry*, 5th ed., VCH, Weinheim, 1985, *A1*, 260.
73. P. Brewbaker, R. Dignan, and S. Meyers, eds, *Singer Clothing Care and Repair*, C. DeCosse, Minnetonka, MN, 1985, *54*, 93–94.
74. L.A. Liddell, *Clothes and Your Appearance*, Good-heart–Willcox, South Holland, IL, 1988, 175, 185.
75. E.O. Wilson, Science, 1998, 279, 2048.

3 The Chemistry of Waste Management and Recycling

3.1 WASTE

The minimization of waste by the chemical industry was discussed in Chapter 1. The waste to be discussed here is municipal solid waste. This is the trash discarded from homes, businesses, construction sites, schools, or other locations. For a sustainable future, it will be necessary to recycle as much of this as possible. The amount of waste varies with the country (Table 3.1), with the United States leading the list as the world's foremost throwaway nation.[1]

The United States generated 208 million tons of municipal solid waste in 1995.[2] This amounted to 4.3 lb./person/day. It consisted of 37.6% paper and paperboard, 15.9% yard trimmings, 9.3% plastics, 8.3% metals, 6.6% food, 6.6% glass, 6.6% wood, and 9.0% of other materials by weight. Containers and packaging accounted for 34.1% of the total. (In 1996, 209 million tons of municipal solid waste was generated, together with 136 million tons of construction and demolition waste.[3]) A study in Clark County, Washington, showed that 2.5% of the material discarded was reusable as it was.[4] The reusable items included food in the original containers, furniture, cosmetics, nuts, bolts, appliances, toys, clothing, carpet, wood building materials (41% of the total), plastic buckets, and ornamental glass. The average business in the United States throws away 250 lb. of paper per person per year.[5] The advent of computers and photocopiers has not reduced the amount of paper used in offices.[6] Seventy percent of the municipal solid waste in the United States ends up in landfills.[7] These have an average remaining life of 10 years.

Only 27% of the municipal solid waste generated in the United States in 1995 was recycled. In 2006, 82 million tons of waste was recycled, which was 32.5%.[8] More than 700 curbside recycling programs and more than 9000 drop-off centers were in operation in 1995. The latter is energy-intensive, because people usually drive to them in their cars. By 2008, there were 8500 curbside recycling programs in the United States.[9] In 1995, the recycled material included 40% of paper and paperboard, 38% of containers and packaging, 52% of aluminum packaging, 54% of steel packaging, 52% of paper and paperboard packaging, 27% of glass packaging, 14% of wood packaging, and 10% of plastic containers and packaging. Office paper was recycled at the rate of 44%, magazines at 28%, clothing at 16%, tires at 17%, nonferrous metals at 70% (largely owing to recovery of lead batteries), and ferrous metals at 31%. (By 1997, the recycling rate for paper and paperboard had grown to 45%.[10]) The United States exports 10 million tons of scrap iron each

TABLE 3.1
Amount of Waste Generated by Various Countries

Country	Waste per Person per Year (tons)
United States	0.88
Australia	0.74
Canada	0.50
Denmark	0.40
Japan	0.35
Germany	0.34
Switzerland	0.32
Sweden	0.28
France	0.26
United Kingdom	0.25
Italy	0.24
Spain	0.22

year.[11] It also exports a million containers of used clothing to developing nations each year.

A model study on Long Island, New York, a few years ago demonstrated that it was possible to recycle 84% of municipal solid waste. It is instructive to look at what some industries have done to approach zero waste.[12] Mad River Brewing Co. in Blue Lakes, California, has a 97% recycling rate. In addition to recycling the standard commodities, such as cans and bottles, paper is printed on both sides, office paper is shredded for packing material, spent grain is sold to local composting companies and farmers, employees repair and use wooden pallets, and wood scraps become firewood. The former Chrysler auto assembly plant in Newark, Delaware, recycled 95% of waste plastic and cardboard, 85% of office paper, 100% of scrap metal, and 20% of wooden pallets.[13] In one year, this avoided disposal costs of US$349,760 for hauling and disposal at a landfill and earned US$212,288 when sold. It should be possible to reuse more of the wooden pallets or to replace them with more durable, reusable plastic pallets.

3.2 RECYCLING

3.2.1 INTRODUCTION

Recycling[14] is simplified if the material is free of contaminants. This requires as much separation as possible at the source. The process is complicated because many consumer items are made of more than one material. A steel beverage may have an aluminum top. The aseptic drink box consists of several layers. Polypropylene film for foods is usually coated with a barrier of poly(vinylidene chloride). It must be remembered that the paper, steel, aluminum, glass, plastic, and such that need to be recycled come in many different grades and compositions. These have been optimized for specific uses, and they may not do well in others. Thus, recycling may have to be to a less-demanding use. Paper can be made by heating and grinding (thermomechanical pulp) or by treating the wood with chemicals to remove the lignin.[15] It may or may not be bleached. Various fillers, sizes, wet strength resins, and other additives may be used on it. Some paper may be dyed or waxed. Steel, aluminum, copper, and other metals are available in many different alloys. Plain steel and stainless steel would not be recycled together. Many different alloys are used in stainless steel. A different alloy is used in the lid (e.g., 0.35% Mn and 4.5% Mg) of the aluminum can than in the body (e.g., 1.1% Mn and 11% Mg) of the can.[16] Deep drawing is important for the body and ease of working the pull tab is important in the lid. Aluminum pans are made of other alloys. Some scrap dealers will not even take aluminum pans. (Aluminum is also alloyed for various uses with iron, copper, silicon, lithium, chromium, lead, zinc, and zirconium.) Window glass has a different composition than that in containers and must be kept separate. Pyrex glass has yet another composition.

The composition of plastics can vary even more. Various polyethylenes vary in molecular weight, molecular weight distribution, degree of branching, length of branching, crystallinity, and so on. For polypropylene, variables of tacticity and crystallinity must be included. Copolymers can be random, block, or graft, and many sometimes contain some of the homopolymers. The additive packages used in the various polymers may vary. These include stabilizers, impact modifiers, flame retardants, and dyes, among others. Coatings and laminates can cause additional problems. Clothing often consists of mixtures of fibers, as in cotton–polyester fabrics. Automobile tires may have one rubber in the tread, another in the carcass, and a polyisobutylene liner inside as a gas barrier. In addition, the degree of cross-linking in rubbers and other polymers may vary.

Some curbside recycling programs allow the comingling of recyclable materials that are relatively easy to separate to save work on the part of the consumer. For example, steel cans can be separated from aluminum cans with a magnet. Separation of a mixture of containers is more difficult, but can be done. If the containers can be passed down a line one at a time, online, and then sensed by infrared, visible, or ultraviolet light, by x-ray, or by other rapid means, blasts of air can remove them into the proper bins. Carpets and plastics can be separated with a handheld spectrometer.[17] Metals and their alloys can be separated with a handheld x-ray fluorescence analyzer.[18] Separation of mixed municipal solid waste, as done in Delaware in past years, has been difficult.[19]

The recycling techniques applied to various clean streams of materials will now be covered.

3.2.2 PAPER

3.2.2.1 Recycling versus Incineration

The recycling of paper[20] has advantages over incineration to recover energy from it,[21] despite claims to the contrary by some authors.[22] Recycling saves more energy than that obtained by incineration, because making new paper requires energy to harvest trees and so on. Recycling creates three times as many jobs as incineration. In England, at least, recycling helps the balance of payments and avoids landfill costs. Those who advocate burning rather than recycling assume that fossil fuels will be used to harvest and pulp the trees. Fossil energy may be used, but it will not have to be in the future. People who recycle paper are working toward a goal of 100% recycled fiber content, with zero wastewater discharge from the plant. In a few instances, this is possible today. Containerboard is being made in Germany[23] and corrugated cardboard in New York[24] from 100% postconsumer material. Newsprint and tissue are being made in a German mill from mixed newspaper and magazines.[25] As the techniques of recycling have gradually improved, the postconsumer content of recycled paper is climbing.

3.2.2.2 Deinking

Some paper can be recycled without deinking. The resulting sheet will be somewhat gray, but can still be reprinted in a legible fashion. For some uses, such as toilet tissue, there should be no need for deinking or bleaching. Some recycled fiber that has not been deinked can be added with a surfactant in the making of new newsprint.[26] Usually deinking is required.[27] Not all uses require the same level of deinking. Sorted office paper can be converted to tissue, white linerboard, and other grades (containing 100% recycled fiber), with less stringent optical requirements than printing and writing grades.[28] The paper is usually repulped in the presence of a base, a surfactant, and other additives at 45°C–60°C for 4–60 minutes. A typical run may contain 0.8%–1.5% sodium hydroxide, 1%–3% sodium silicate, 0.25%–1.5% surfactant, 0.5%–2.0% hydrogen peroxide, and 0.15%–0.4% diethylenetriaminepentaacetic. The fiber is swollen by the sodium hydroxide at pH 8–10, reducing the adhesion of ink to the fibers. The sodium silicate prevents the redeposition of ink, and the hydrogen peroxide counteracts yellowing. The diethylenetriaminepentaacetic acid ties up any metal ions present that might cause the hydrogen peroxide to decompose. The ink[29] particles are then removed by washing the pulp or by flotation.[30] Washing is best for small particles of ink and flotation for larger particles.[31] The presence of clay filler particles from magazines improves the efficiency of flotation deinking, so that newspapers and magazines are often repulped together.[32] Pressurized flotation cells are better than those operating at atmospheric pressure.[33] Column flotation is superior to that done in a tank.[34] In some flotation deinking, calcium chloride is used with a long-chain fatty acid to make the ink particles hydrophobic.[35]

Mechanical means are also used to enhance the removal of ink from the fibers. Passing the fibers between rotating disks can help.[36] When ultrasound is used, there is less need for chemicals, and the resultant pulp is stronger.[37] Steam explosion, during which paper containing 50% moisture is passed through a hot coaxial feeder, reduces the need for chemicals and eliminates the need for flotation.[38] This process can also handle laser and xerographic inks, which can be difficult to remove by other means. There is some loss in tensile strength.

3.2.2.3 Enzymatic Repulping

No deinking chemicals are necessary when old newspapers are repulped with a cellulase.[39] The larger uninked fibers are retained by a plastic mesh screen, whereas the smaller inked ones go through to a separate vessel. Enzymes facilitate toner removal, increase brightness, improve pulp drainage, preserve fiber integrity, and lower chemical costs in the repulping of mixed office wastepaper.[40] The drainage is best when the enzyme is combined with a polyacrylamide containing quaternary ammonium groups.[41] Xylanases improved drainability without hurting tensile strength when repulping old paperboard containers.[42] Buckman Labs markets an esterase to cleave esters in adhesives on wastepaper during repulping.[43] The deinked pulp can be made into fine writing and printing papers. Old newspapers have been converted to supersoft tissue using enzymatic repulping.[44] Enzymes should be useful in removing soy-based inks.

3.2.2.4 Extent of Recycling of Paper

Europe recovered 48 million tons of paper, nearly 50% of the total used, in 2001, of which 40 million tons went back into paper and paperboard.[45]

Some mills have been able to obtain very high rates of recycle of postconsumer paper. U.K. Paper uses deinked pulp of 75%–100% recycled content.[46] Bridgewater Paper, in England, produces newsprint that averages 95% recycled content.[47] Some urban minimills[48] can make recycled linerboard from 100% old corrugated containers.[49] A mill next to the former Fresh Kills landfill on Staten Island, New York, makes 100% recycled containerboard.[50] It "harvests an urban forest." New Jersey had 13 paper mills running only on wastepaper and eight steel minimills using scrap steel in 1999. The total annual output of these two industries was more than US$1 billion.[51] Some of these minimills are able to operate closed systems with no effluent.[52] Zero liquid effluent is now practical for corrugating medium and linerboard mills.[53] A few other mills have achieved it.[54] A Kimberly–Clark mill in Australia will pulp thinnings from *Pinus radiata* plantations with magnesium bisulfite, bleach with hydrogen peroxide, recover all spent sulfite liquor, capture all gases from cooking, and either compost all fines or use them for fuel.[55] Further study of the use of reverse osmosis, ultrafiltration, biological oxidation, and such may make it possible for other mills to have zero effluents. Paper mill effluents have been notorious for their contamination of streams.[56] Even the natural resin and fatty acids in the wood can be toxic to fish.[57] It would be better to recover these acids for industrial use, as done in many mills.

3.2.3 PLASTICS

The recycling of plastics[58] is not growing rapidly. In the United States, less than 10% of plastics packaging is recycled. Only 4% of plastic grocery bags are recycled (e.g., to decking lumber).[59] In 1993, only about 2% of all plastics produced were recycled.[60] Some recycling facilities are being sold to smaller companies by the large chemical companies or are closing.[61] One problem is obtaining an assured stream of clean material to recycle. The few success stories include poly(ethylene terephthalate) bottles, many of which are returned under deposit legislation.[62] In 1996, 485,000 metric tons was collected in the world for recycling. (Of this, 330,000 metric tons was in North America.) There is even a report of a shortage of these bottles for recycling.[63] These are used to produce foamed insulation,[64] fibers for various applications, shower curtains, paint brushes, and others. However, the recycling rate for polyethylene terephthalate bottles dropped from 27.8% in 1996 to 25.4% in 1997 to 16% in 2007, whereas total use was rising.[65] The recycling rate for high-density polyethylene was 6% in 2007. During the same time period, the recycling rate for all plastic containers dropped from 24.5% to 23.7%. In 2005, 2.1 billion lb. of plastic bottles were recycled in the United States, which was 24.6% of the total.[66] This includes both poly(ethylene terephthalate) and high-density polyethylene bottles, which made up 96% of all plastic bottles. Small amounts of poly(vinyl chloride) with the polyester can seriously contaminate the batch.[67]

High-density polyethylene milk jugs are being recycled into other containers, usually for nonfood uses, as well as in drainage pipes, trash cans, grocery bags, traffic barrier cones, and such. If they go into new food containers, they must be in an inner layer in a laminate with virgin material on the surface. The polypropylene cases of car batteries are also recycled well. Although poly(vinyl chloride) can be recycled into bottles, pipes, and window frames, it is recycled less than other plastics.[68] Solvay recycles poly(vinyl chloride) in plants in Italy and Japan.[69] Ground plastic is dissolved in methyl ethyl ketone at 100°C–140°C, the solution is filtered, and then the polymer is precipitated. Metals, poly(ethylene terephthalate), and nylon do not dissolve.

The plastics that are in other throwaway packaging[70] are difficult to handle. They may be in thin films or contaminated with food residues, and may be in a great many trash cans in a great many places. The high costs of collecting and sorting raise the prices of recycled plastics to a point that most manufacturers would rather use virgin plastics. (The cost of virgin resin does not include global warming and natural resource depletion.) If the pieces of plastic can be sent down a line one at a time, an online spectroscopic analyzer can sort them. For example, polyethylene and polypropylene can be separated by pulsed laser photoacoustic technology.[71] Handheld devices that work by Raman spectroscopy are also available.[72] If the materials to be sorted are crumpled films, laminates, core-sheath fibers,[73] or polymer blends, such systems will not work well. Foamed cups and "peanuts" of polystyrene are voluminous and require some form of densification to make transportation to the recycling center economical.[74]

Many plastic objects are marked with a number inside three chasing arrows: 1 is polyethylene terephthalate; 2 is high-density polyethylene; 3 is poly(vinyl chloride); 4 is low-density polyethylene; 5 is polypropylene; 6 is polystyrene; and 7 is other (including a multilayer container). Such a system allows a consumer to place the used object in the proper bin, if such bins are provided.

3.2.3.1 Recycling Methods

Recycling is easiest when only one polymer is present and no separation is needed. Polypropylene fertilizer bags can be reprocessed without additional stabilizers if they were properly stabilized the first time.[75] Because of the degradation in aging and remolding, it is often necessary to add additional stabilizers when a plastic is recycled into a new plastic object.[76] Carpets are often made with one polymer in the facing and another in the backing. Hoechst Celanese has devised one made entirely from polyester that should be easier to recycle at the end of its useful life.[77] This also simplifies what one is to do with scraps from the manufacturing and installation processes.

The poly(ethylene terephthalate) soda bottle is a mixture in that it has a label and cap of polypropylene. Redesign has eliminated the polyethylene base cup used formerly. It may be possible to prepare an all-polyester soda bottle.[78] The label might be replaced by inkjet printing. A shrink label of polyester is also possible, but care would be required in its installation because its softening point is about 100°C higher than that of the current label of polypropylene. A cap of polyester may also be possible if crystallization of the polymer can be sped up to shorten molding cycles. Sodium bicarbonate and carbonate have been used as nucleating agents to shorten the molding cycles of recycled poly(ethylene terephthalate).[79] (Bottle makers often add a crystallization inhibitor so that the bottles remain clear. A cap would not have to be clear.) The seal of the cap liner might have to be made of a different polymer, which in the small amount involved, one hopes, would not interfere with recycling of the bottles. DuPont is marketing a glass fiber-reinforced molding compound made from used soda bottles and x-ray film from which the emulsion has been removed.[80] The key to this reuse is to prevent degradation of the molecular weight by keeping the polyester dry to prevent hydrolysis, minimizing oxidative degradation during recycle, and keeping rogue materials, such as polyethylene and poly(vinyl chloride), out of the system.[81]

3.2.3.2 Use of Compatibilizers and Handling of Mixtures

Much effort is being expended on polymer blends to reduce the need for separation and to improve the properties of recycled materials. Because most polymers are incompatible with one another, this usually requires the addition of a compatibilizer[82] or the preparation of one *in situ*. (Compatibilizers are often block or graft copolymers, with parts similar to each of the polymers that are to be made compatible.) A polypropylene grafted with maleic anhydride has been used to compatibilize blends of polypropylene and nylon-6,6.[83] (The amino end groups in the polyamide react with the anhydride groups in the graft copolymer to form a new amide.) A mixture of poly(ethylene terephthalate) and linear low-density polyethylene can be compatibilized using 10% of the sodium salt of poly(ethylene-*co*-methacrylic acid).[84] Perhaps, this could be used on a mixture of soda bottles and milk jugs. Reactive extrusion of a mixture of poly(vinyl chloride) and polyethylene with a peroxide and triallylisocyanurate gave a compatible mixture.[85] In this case, the compatibilizer was formed *in situ* by grafting. The product was a cross-linked copolymer. Moldings of a mixed waste of polystyrene and polyethylene had increased strength.[86] For this, the mixture was not compatible. The improvement in properties was due to the separation of the less than 30% polystyrene as fibers. Small amounts of hydrogenated styrene–1,3-butadiene block copolymers were used to compatibilize mixtures of polystyrene and polyethylene.[87] Blends of polypropylene and polystyrene were compatibilized with 2.5% polystyrene-block-poly(ethylene-*co*-propylene).[88]

3.2.3.3 Use of Chemical Reactions in Recycling[89]

The concern about possible contamination of food containers means that they can be recycled to only nonfood contact uses in the United States. To reuse poly(ethylene terephthalate) in contact with food, it must be broken down, purified (to remove metal compounds, colors, and such), and resynthesized (Schematic 3.1). This has been done by hydrolysis, methanolysis, and glycolysis[90] (The solvolysis can be complete in 4–10

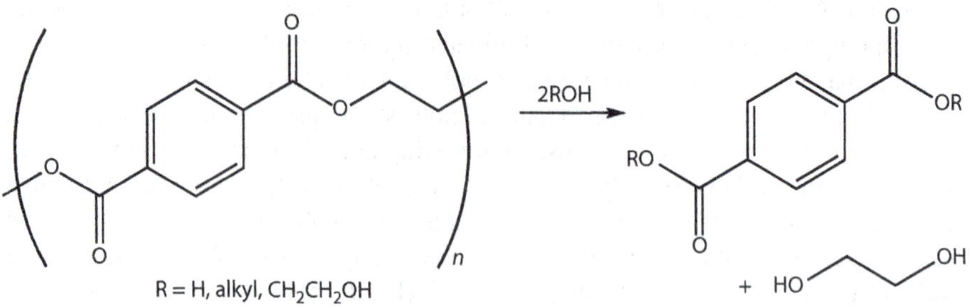

R = H, alkyl, CH$_2$CH$_2$OH

SCHEMATIC 3.1 Breakdown of poly(ethyleneterephthalate)

minutes when microwaves are used with a zinc acetate catalyst.[91] It can be done in 10 minutes at 380°C.[92] For nonfood uses, such as the use of poly(ethylene terephthalate) in magnetic tapes, it may be sufficient to melt the polymer and filter out the chromium or iron oxides.[93])

The technique is also applicable to cross-linked polyesters made from unsaturated polyesters and styrene, as in the sheet-molding compounds used in cars and boats.[94] The ester linkages were broken by treatment with ethanolic potassium hydroxide or by ethanolamine. The latter was preferred, because no waste salts were formed by neutralization, as they were with the potassium hydroxide. The filler and glass fibers could be removed at this stage, if desired. The alcohol-soluble product was suitable for use in new bulk-molding compounds. A method that allows the cross-linked polyester to recycle back into the same use involves heating with propylene glycol, re-esterification with additional maleic anhydride, and adding more styrene.[95] An alternative way of recycling the sheet-molding compound would be to pulverize it and use it as a filler in new sheet-molding compound. Waste polycarbonates can be broken down by phenolysis.[96] The bisphenol in it can be recovered in 80%–90% yield by methanolysis[97] or hydrolysis.[98] (Direct reuse by remolding would be preferable when possible.) Polytetrahydrofuran can be converted to monomer by a hot treatment with hydrochloric acid on kaolin.[99] The solid catalyst was still active after repeated use.

3.2.3.4 Biodegradable Polymers

Biodegradable polymers[100] have been advocated by some to solve litter problems and to offer composting as a way of disposing of waste plastic. Merely being in a landfill will not ensure the decomposition of plastic. During the excavation of landfills, layers have been dated by the dates on recovered newspapers.[101] Synthetic polymers such as polyethylene were made to degrade in sunlight or fall apart by bacterial action in several ways.[102] One is to put some carbonyl groups into polyethylene by including a vinyl ketone in the polymerization. Another is to accelerate photodegradation by the inclusion of a metal salt [e.g., an iron(III) dithiocarbamate], which may be coated with a material that has to be removed by biodegradation before the photodegradation begins. Time control is obtained by an appropriate choice of metal compound and its concentration. Starch-filled polyolefins are degradable in that the bacteria can remove the starch, but the polyolefin itself remains. (In the United States the law requires that the plastic rings that hold six beverage cans together must degrade in sunlight to avoid animals becoming caught and dying in the rings.)

Biodegradable polymers certainly have a place in the future, but with some limitations. If they are on paper that is being recycled, they will have to be designed to come off in the process. If they get into streams of plastics being recycled, the recycling could be ruined. For example, starch would decompose below the molding temperatures of nylon or polyester. If photodegradants enter the waste plastic stream by this route, the lifetime of the remolded plastic may be short. It will be difficult to devise a sorting and labeling system that will keep the usual synthetic plastic and biodegradable polymers separate in the waste stream. (A school system in Virginia decided to use biodegradable cutlery and trash bags so that they could be composted together with food waste.[103]) The best approach may be to cut back on the use of throwaway items in favor of those that are durable and last a long time in repeated use.[104]

3.2.4 METALS

3.2.4.1 Methods of Recycling

Aluminum, lead, and steel are the principal metals that are recycled.[105] The aluminum comes mostly from beverage cans. Of the 130 billion metal cans used in the United States each year,[106] 99 billion are made of aluminum.[107] In 1997, 59.1% of the aluminum beverage cans made in the United States were recycled.[108] In 2007, 54 billion cans were recycled, 53.8% of the total.[109] Americans discarded 32 billion soda cans containing 435,000 t of aluminum in 2002.[110] Cars are the largest single source of scrap iron.[111] The lead is from car batteries. (In the United States, 96.5% of lead–acid batteries were recycled in 1996.[112]) Clean streams of metals recycle better. Non-can aluminum consists of different alloys than the cans and should not be mixed with them for recycling.

Physical methods of separation are preferred to chemical ones.[113] Magnets are used to separate steel cans from municipal solid waste. After the battery, gas tank, radiator, and reusable parts are removed from a used automobile, it is shredded. Air classification removes the fabrics and light plastics. The steel is taken out by a magnet. The aluminum is separated from the copper, zinc, and lead by density using a sink–float technique and then from the remaining materials in an eddy current separator. If the polyurethane foam, which can be recycled,[114] is removed mechanically, the remaining plastics can be separated by dissolution in various solvents, acetone for acrylonitrile–1,3-butadiene-styrene terpolymers, tetrahydrofuran for poly(vinyl chloride), and xylene or cyclohexanone for polypropylene. Much thought has been given to the preparation of cars that will be easier to recycle at the end of their useful lives.[115] DuPont has made a prototype "greener" car in which poly(ethylene terephthalate) is used to replace many of the other plastics found in the usual car. Polyester cushions, seat covers, and headrests were used. DuPont says that the polyester part could be recycled as a unit or remolded in the same way that soda bottles are.

Steel food containers should be detinned before melting down to recover the valuable tin and to avoid contamination of the batch of steel. Impurity zinc can be removed from aluminum by distilling it out at 900°C under argon in 1 minute.[116] Zinc can be recovered from galvanized steel by distillation at 397°C at 10 Pa.[117] It can also be recovered in 99% yield by extraction with 40% aqueous sodium hydroxide at 200°F, followed by electrodeposition of the zinc, which recycles the sodium hydroxide to the process.[118]

3.2.4.2 Recovery of Metals from Petroleum Residues

Nickel and vanadium in the fly ash from the burning of heavy oil can be recovered by chlorination and distillation, with 67% recovery of the nickel and 100% recovery of the vanadium.[119] Another method is to leach out the nickel with aqueous ammonium chloride, followed by treatment with hydrogen sulfide to recover nickel sulfide (in 87% yield), which could be refined in the usual way.[120] The vanadium was recovered next in 78% yield by solvent extraction with trioctylamine, followed by treatment of the extract with aqueous sodium carbonate to take the vanadium back into water, and finally ammonium chloride to produce ammonium vanadate.

3.2.5 GLASS

Most glass recycling[121] is done with jars and bottles that are melted down and made into more jars and bottles. As with metals, this process can be done over and over again. Other types of glass, such as windows and light bulbs, as well as ceramic tableware, have different chemical compositions and must be kept out of the container-making process. They too could be recycled into more window glass and light bulbs, if there were enough of them in one place at any given time. The containers must be sorted by color (clear, green, and brown). If the containers are whole, this can be done automatically by an online color sensor. Broken glass of mixed colors is a problem, especially if it is in small fragments, and might have to be diverted to another secondary use. After sorting by color, the glass is pulverized. Paper labels can be removed by screening or suction. Steel caps are removed with a magnet. The pulverized glass (called cullet) can be melted down and made into new containers. More commonly, additional sand, limestone, and sodium carbonate are added in making a new batch of glass. The addition of the cullet lowers the melting point, saves energy, and prolongs furnace life.

3.2.6 MISCELLANEOUS RECYCLING

3.2.6.1 Composting

Yard wastes (grass, leaves, and tree limbs) constitute 15.9% of municipal solid waste, which can easily be recycled by composting. The best method is to eliminate the grass clippings by cutting the lawn frequently and leaving them on the lawn where they fall. The leaves can be composted on site, in the backyard, which eliminates the cost of collection. With the proper moistening and the turning of the pile, the odor will not be a problem. Food wastes can be included if the composter has a screen to keep out dogs, raccoons, rats, and such.[122] Under proper composting conditions, the decomposition process will produce temperatures of 56°C–60°C, which will kill any pathogens present. Composting can also be done on a neighborhood or on a municipal basis, although there will be a cost for collection.[123] In Newark, Delaware, leaves are raked to the curbside by the homeowner and then sucked into a city truck for transport to the municipal composting facility. The resulting compost is available to residents of the city. Many states and cities now prohibit yard waste in the trash. Texas has 50 large-scale composting operations.[124] Brush disposal in Texas landfills has dropped 75% since 1992 owing to this. More than 8000 American farms now have composting operations.[125] African cities are also composting.[126]

3.2.6.2 Uses for Food-Processing Wastes

Food-processing wastes are often used as food for animals. Corn gluten meal, left from the preparation of high-fructose corn syrup, can be used as food for fish in aquaculture.[127] So can the distillers dry grains left from the fermentation to produce ethanol. Dietary fiber for consumption by humans can be recovered from pear and kiwi wastes.[128] Tomato skins can be removed by steam and abrasion in a process that eliminates the traditional use of sodium hydroxide for peeling. The waste can be fed to animals. Brewer's spent grain has been used to make bread.[129] A plant in Missouri converts turkey scraps into an oil resembling biodiesel at 100–200 barrels/day.[130]

3.2.6.3 Toner Cartridges

Some manufacturers have taken back used toner cartridges (used in laser printing), cleaned and inspected them, replaced worn parts, if needed, refilled them with toner, and sold them.[131]

3.2.6.4 Use of Baths of Molten Metal and Plasma Arcs

When no other way of recycling is available, organic wastes can be dropped into a bath of molten iron, at 1650°C, to produce carbon monoxide, hydrogen, and hydrogen chloride (if chlorine is present).[132] Lime and silica fluxes may be used to form a slag layer. Heavy metals collect in the bath. No nitrogen oxides, sulfur oxides, or dioxins are evolved. Destruction of organic compounds is more than 99.999%. The synthesis gas formed can be used to make methanol or for hydroformylation. Any hydrogen chloride formed would have to be removed by a scrubber. This is said to be cheaper than incineration.

A plasma arc torch, operating at 3000°F, converts wastes to hydrogen, carbon monoxide, metal, and a glassy slag, so that nothing is wasted and nothing goes to the landfill.[133] The slag can be used in asphalt or concrete. Air emissions are one tenth those of an incinerator. The process could make landfills obsolete. Fort Pierce, Florida, planned to build a US$425 million plant to process 8000 t of municipal solid waste per day plus what is in the existing landfill but has scaled back the size of the plant.[134] Two plants are operating in Japan. Others are in Washington state and Hawaii. Startech Environmental Corporation markets the technology.[135] Aseptic juice containers consist of layers of paper and aluminum plus adhesive layers. Tetra Pak and Alcoa recycle these with plasma jet technology in Brazil.[136]

3.2.6.5 Uses for Inorganic Wastes (Other than Glass and Metals)

A few examples will be given of efforts to recycle inorganic wastes. Only 22% of coal fly ash is utilized, with outlets in cement and concrete products, filler in asphalt, grit for snow and ice control, road base stabilization, and others.[137] Concrete made with fly ash is stronger and is up to 30%–50% cheaper.[138] Building blocks can be made from fly ash, waste calcium sulfate from flue gas desulfurization, and lime or Portland cement.[139] More could be used if we built our houses of concrete blocks instead of wood. Efforts have also been made to recover the aluminum, iron, and other metals in fly ash. This would reduce the need to mine so much new ore. As the world switches from fossil fuels to renewable sources of energy, the problem of what to do with fly ash will disappear. The ash from incinerated sewage sludge has been mixed with silica, alumina, and lime for conversion to tiles and floor panels in Japan.[140] Spent bleaching earth from the treatment of edible fats and oils presents a disposal problem.[141] The best approach is to regenerate it for reuse. Extraction with isopropyl alcohol or hot water under pressure removes adhering oil so that the solid can be reused as a filter cake. It can also be put into a cement kiln or used in land application. Contaminated sulfuric acid can be cleaned up by cracking at 1000°C, followed by conversion of the exit gas to fresh sulfuric acid.[142] The process is in use at three plants in Europe and at six in the United States. Zero discharge of wastewater is the goal of industry. This involves biological treatments, reverse osmosis, and so on. The color of spent dyebath water can be removed by treatment with ozone so that the water can be reused.

3.3 METHODS AND INCENTIVES FOR SOURCE REDUCTION

3.3.1 RANGE OF APPROACHES

The following hierarchy of methods of use goes from those with the most waste to those producing the least waste. (An example is given in most cases.)

Use once and then bury it in a landfill (throwaway packaging).

Use once and then compost it (biodegradable packaging).

Use once and then burn it with recovery of energy (throwaway packaging).

Use once and then make it into a different object, which after use can be burned or buried (high-density polyethylene milk jug made into a container for motor oil).

Use once and then make it into a different object that will last for many years (a polyethylene terephthalate soft-drink bottle made into fiber for clothing or a sleeping bag).

Use once, and then melt and reshape it to original use (recycled aluminum cans or single-use glass bottles).

Use once, and then melt, and reshape it to a different long-term use (steel can be made into a reinforcing bar for concrete).

Use once and then hydrolyze the polymer to recover the raw materials, which are then converted into a duplicate of the original object (a polyethylene terephthalate soft-drink bottle made into another soft-drink bottle).

Use 10–60 times and then remake into a duplicate object [a refillable glass or a poly(ethylene terephthalate) soft-drink bottle].

Use 10–60 times and then remake into a different object with a long life (a polycarbonate milk bottle recycled into a part for an automobile).

Use repeatedly, sharpen as needed, and replace small parts that wear out (saw and safety razor).

Use indefinitely (China mug or drinking glass).

Is the object or its content needed at all (bottled water, neckties)?

3.3.2 THROWAWAY ITEMS AND THE CONSUMER

Source reduction for the consumer means obtaining products that he or she needs with a minimum of waste to be discarded. The most important approach is to substitute reusable objects for single-use, throwaway ones. It is instructive to compare what is done today in the United States with what was done 50 years ago and with what is still done in many other countries (Table 3.2).[143]

TABLE 3.2

Approach to Substitution of Reusable Objects for Throwaways

Today	What Was Formerly Done
Paper napkin	Cloth napkin
Paper tissue	Handkerchief
Plastic bag for waste	Wash out garbage can
Discard broken shovel and buy a new one	Replace broken handle on shovel
Single-use soft-drink container	Glass bottle with US$0.02 deposit
Single-use polyethylene milk jug	Milk delivered to door in refillable glass bottles
Single-use glass cider jug	Take own bottle to farm for cider
Assortment of two dozen screws in plastic container	Buy one screw at local hardware store
Discard faulty toaster	Local repair of toaster
Two disposable mouse traps in plastic package	Buy one mouse trap and use it for years
Bottled water	Canteen
Disposable polystyrene cup	China cup or drinking glass
Disposable diapers	Cloth diapers collected from home and cleaned by diaper service
Disposable camera	Use camera indefinitely, just buy new film as needed
Disposable razor	Safety razor used indefinitely, just buy new blades
Aluminum pan under frozen TV dinner	Use own pan repeatedly
Cling wrap and aluminum foil to wrap food	Wax paper
Plastic grocery bags	Paper grocery bags or market basket
Weed whacker or herbicide	Hand clippers
Recreation vehicle	Tent
Disposable needles for syringes	Sterilize needles in autoclave
Electric pencil sharpener	Turn sharpener by hand
Photocopier	Copy by hand or blueprint
One car per driver	Ride bus or street car
Disposable ballpoint pen	Fountain pen
Leaf blower	Rake

Too many items are discarded because a person wants something new; a fad passes and an item goes out of fashion; the person gains or loses weight; the child outgrows it; it was stored improperly; a minor repair or refinishing is needed; and so on. It should be possible to reverse some of these trends to minimize waste, energy consumption, and natural resource depletion without adversely affecting the health or longevity of people. One third of Americans today are obese, in part because of all of the "labor-saving" devices to which they have become accustomed.

There is some question over why some of the items in Table 3.2 need to be disposable. Two billion disposable razors are used each year in the United States. Reusable safety razors work, with the need to dispose of only used blades. This author has minimized his use of razors and shaving cream, while saving 10 minutes every day, by wearing a beard. Kodak advertises its single-use camera as 86% recyclable. A regular camera lasts so long that recycling is a minor issue. It requires only new film. No film is needed in digital cameras. The use of wooden pencils has resulted in overexploitation of red cedar (*Juniperus virginiana*) trees and more recently of *Libocedrus decurrens*.[144] Pencil board can be made today from recycled paper.[145] A mechanical pencil requires only replacement lead, which is used completely. A fountain pen requires only replacement ink. A cutout hole eliminates the need for a plastic window in an envelope. A rubber stamp or a fountain pen eliminates the need for an adhesive return address label. A moist sponge can eliminate the need for the strip of adhesive tape sometimes used to seal envelopes. If the material inside is not personal, the flap can just be tucked in. An air-cooled engine requires no engine coolant.

3.3.3 CONTAINERS AND PACKAGING

Containers and packaging[146] constitute 34% of municipal solid waste by volume. There are many ways in which this can be reduced without lowering the standards of sanitation. More businesses are shifting to multitrip containers for delivering their products. In one case (Nalco), the use of 96,000 15–800 gallons stainless steel containers has eliminated the need to dispose of 3 million drums and 30 million lb. of chemical waste since the program began. Use of returnable, refillable, pesticide containers reduces the likelihood of contamination of the environment, while saving time and money.[149] Manufacturers ship more than 90% of all products in the United States in corrugated cardboard boxes.[150] Multitrip containers for the delivery of food to supermarkets could save large amounts of corrugated cardboard. This calls for some innovation to produce easy-to-open and -close containers that nest or can be collapsed for reshipment to the manufacturer. Standardization of package sizes or at least crate sizes could allow them to be used by many different companies, with credit going to the company that provided them in each specific case. A German manufacturer is studying the use of collapsible polypropylene crates for use with its products. These might be made of plastic or other material. 3M has a reusable, collapsible-packaging system that saves the company US$4 million/year.[151] Frigidaire has a similar program.[152]

There are also many opportunities for reduction of packaging at the consumer level. The Giant chain of supermarkets in the eastern United States offers over 100 different foods and other items in bulk (in addition to the usual fruits and vegetables). The customer uses a scoop to fill a plastic bag, and then puts a sticker and a plastic tie on it. The items include cereals, candy, dried fruits, nuts, pasta, snacks, and pet supplies. They are usually cheaper than those that are prepackaged. Unfortunately, this is only a small portion of the items sold in the store. It should be possible to add beans, detergents, coffee, and other items. Beverages might be dispensed into the customer's own container, as done with milk in parts of Germany. The only requirement may be that the cashier at the checkout counter be able to recognize the item inside the bag or jar. Cereal in large boxes, instead of in individual variety packs, is 78% less expensive and generates 54% less waste.[153] Buying fruit juice in frozen concentrate, rather than in shrink-wrapped packs, cuts the cost by 41% and the waste by 61%. The least packaging for breakfast cereal uses the form-and-fill package made of plastic film and uses no outer box. The box of rolled oats used to be made of all paper. Now it comes with a plastic top, which makes it harder to recycle. Some bread is double wrapped, even though a single layer is sufficient. Some fabric softeners are now available as liquid concentrates. The customer buys the original container, which can then be refilled with concentrates as needed. There are also examples in which small items are put into oversized packages so that the package can carry an advertisement. It would be better to put the advertisement in the bin where they are kept.

Milk and orange juice are also sold in half-gallon bisphenol A polycarbonate jugs. A jug may cost US$1.10, but the cost per trip is low when spread over 40–60 trips.[154] A deposit of US$0.25 ensures a high return rate. Some 8 fl oz. school milk bottles of polycarbonate realize over 100 trips. Just as with the polyester bottles, these also have to be washed under mild conditions to avoid hydrolysis. (Bisphenol A is an endocrine disrupter, which may be detrimental to the health of fetuses and young children.) At the end of their useful life, the manufacturer (formerly General Electric, now SABIC) takes back the bottles for remolding into bottle crates and such. This system has been used in the United States, Sweden, and Switzerland for over 10 years. Most milk is sold in paper cartons coated with polyethylene, or in jugs made of high-density polyethylene. The system for milk requiring the least packaging is probably powdered skim milk equivalent to 22.7 L (24 qt) of liquid milk in a cardboard box covered with aluminum foil. It can be stored at room temperature and requires refrigeration only after it is made up as liquid milk. The polycarbonate bottle cannot be used for soft drinks because it is not a suitable barrier to carbon dioxide. A challenge is to find a suitable barrier that will not interfere with recycling the container at the end of its useful life as a bottle. An inorganic barrier coating, such as silica, might work. — TASK

Bottled water is popular in the United States today. Some people feel that it tastes better, has fewer impurities, and may confer higher social status than tap water.[155] The analysis of 37 brands in 1991 found that 24 had one or more values that were not in compliance with the drinking water standards of the United States.[156] Some were just bottled municipal water. Testing of 1000 bottles of 103 brands in 1999 found that about "one third of the brands had at least one sample containing levels of contamination that exceeded state or industry limits or guidelines."[157] Bottled water does not contain sufficient fluoride ion for good dental health.[158] It would be better to drink at the nearest water fountain or to fill your reusable bottle there. This would eliminate some plastic container waste. (Bottled water does have a place in

[handwritten in margin: How to recycle PET bottles ?]

some of the less-developed countries in which the standards of sanitation may be inadequate to protect the public water supply.) One company uses refillable bottles of poly(ethylene terephthalate) for its mineral water in Holland and Belgium.[159] The headspace of returned bottles is checked by gas chromatography–mass spectrometry at the rate of 0.1 s/bottle. The label is cut off with a high-pressure water jet at the rate of 550 labels/minute. The sleeve waste is recycled. The bottles are cleaned at 167°F, inspected physically, filled, and a new shrink label of polyethylene applied. The crate of six 1.5-L bottles has a deposit of US$7.20, half for the crate and half for the bottles (i.e., US$0.60/bottle). The return rate is 94%. Americans drank 24 gallons/person at a cost of US$26 dollars.[160] This is such a waste of plastics, energy, and dollars that some cities, such as San Francisco, California, have banned their sale.

3.3.4 Using Less Paper

The United States consumes enormous amounts of paper each year, 1.5 t/household/year.[161] It takes 5 billion ft^3 of trees plus 35 million tons of recycled wastepaper to make this paper. Each year Americans receive 4 million tons of junk mail, 47% of which is discarded by the recipient without looking at it.[162] It took 1.5 trees for the junk mail received at each household. Each U.S. office worker discarded 225 lb. of paper in 1988.[163] Various methods have been suggested for reducing office waste, such as using both sides of the page, using smaller sheets for short memos, using e-mail, circulating memos instead of making copies, eliminating coversheets and wide margins, or making fewer photocopies. New incentives are needed for making fewer photocopies and reducing the number of computer printouts. (Charging individuals more for each one might help.) Various predictions of a cashless society, where transactions would be done by computer and e-mail, have been made in the past. This is starting to happen. Use of e-mail for bank statements and other investment information could reduce the volume of mail. If every home with a television set could use it for this purpose, the volume of paper used in mail would be reduced. This may require further developments in technology.

Because the newspaper is such a large component of household paper, it deserves special consideration. It can be replaced by a system that transmits the information in the newspaper to the customer over the telephone lines during the night. In the morning, the customer queries the terminal to obtain the table of contents. He might also print out the portions of interest. When such systems were test-marketed earlier, they flopped. One problem may have been that they cannot be taken to the office to be read at lunchtime. A standard command in the terminal to print out the front page, the sports scores, and the stock prices, or whatever else was of special interest, might have taken care of this. It is also possible that the system was tried before people became used to the World Wide Web and similar systems. Updating the news sections hourly might be a selling point with some customers. Most scientific journals are now online. Popular magazines might be handled in the same way. This would eliminate the need for finding space for them in the home and the need for finding a place to recycle them. It is also possible to read them at a central library, either in person or through a website. Sharing with friends is a good way to use magazines that contain no time-sensitive material.

3.3.5 Life-Cycle Analyses

Life-cycle analyses[164] are becoming increasingly popular. They consider the amounts of energy and materials used, together with the waste and pollution produced, in taking a product from cradle to grave. ("Cradle to cradle" is a better term since it implies the use of as much waste as possible for new useful material.[165]) Unfortunately, they are not all done in the same way or with the same assumptions. Most assume the use of energy from fossil fuels rather than from renewable sources. Very few consider the costs of global warming and natural resource depletion, not to mention the loss of biodiversity. It is important to see what assumptions have been made and who paid for the study, for it may not be one that is completely objective. The use of paper versus polystyrene for cups may tend to favor the foamed polystyrene one unless a China mug or a drinking glass is included.[166] The reusable one will win because washing it is not very expensive and does not involve the energy and materials to make a new one. Paper versus plastic is also debated for grocery bags. The winner here is the string bag or market basket that the customer brings to the store. Some paper bags have a statement on them that they are recyclable and are made from a renewable resource. Some also say, "reuse this bag for 2 cents credit." Apparently, this is not enough of an incentive for customers, for very few of them do it. (Customers could also bring back for reuse the thin polyethylene film bags used to carry and store fruits and vegetables.) A local supermarket has used bags of high-density polyethylene with labels that say, "Save a tree! By using plastic bags, you help save the trees and forests that are cut each year to make paper bags. Recycle! Help the environment." This is encouraging the use of a nonrenewable resource. It also ignores the fact that the recycling of plastic films is not working well. A thin plastic bag closed by heat-sealing, a clip, or a paper or plastic tie might replace the shrink-wrapped polystyrene tray. Trays and egg cartons can be made of paper, a renewable resource, by pulp molding, the way they used to be. One company has been "earning greenie points" by packaging its toaster ovens in a molded pulp made from 100% used newspaper.[167] This replaced foamed polystyrene.

3.3.6 Role of Government in Reducing Consumption

It is apparent that source reduction and higher recycling are unlikely to happen without some form of government intervention.[168] This can take the form of laws banning certain uses, taxing undesired practices, subsidizing desirable practices, labeling laws showing true costs, or other such. Thus, it is possible to consider taxes on throwaway items and those based on nonrenewable raw materials, such as petroleum, natural gas, and coal. Denmark's tax on waste, started in 1987, is the highest in Europe.[169] Austria, Belgium, Finland, France, the Netherlands, and the United Kingdom also tax waste. Sweden and Norway are considering such taxes. The U.S. Congress is studying the removal of the hundreds of billions of dollars of subsidies given each year to virgin resource producers.[170] These subsidies to the mining, timber, and petroleum industries are considered to be impediments to recycling. Maine has a law requiring the payment of disposal fees at the time of purchase of major new appliances, furniture, bathtubs, and mattresses.[171] There are also laws that require a minimum amount of recycled content in new items.[172] Sweden and Germany have banned unsolicited direct mailings.[173] If the U.S. Postal Service would require first-class postage, the amount used on personal letters, for junk mail, the amount of such mail would decrease.

The 65% recycling rate for aluminum beverage containers in the United States in 1994 (54% in 2007) is considered high.[174] However, if one looks at the billions of cans that are lost from the recycling loop each year, it is easy to see that a better system is needed. There is a proven way to obtain more of them back. Saphire has reviewed the role of deposits in retrieving used beverage containers.[175] Maine redemption rates for beer and soft-drink containers are 92%, for wine containers 79%, and for juices 75%. In the other 10 states with deposit-refund laws, only beer and soft-drink containers are covered. New York has the lowest redemption rate of the states with bottle bills, with a redemption rate of 66% for soft drinks and 79% for beer. Delaware's bottle bill is relatively ineffective, because aluminum containers and 2-L plastic containers are exempted. British Columbia is considering expansion of its 27-year-old bottle bill to cover sports drinks, bottled water, fruit and vegetable juice, wine, and spirits.[176] A nationwide bottle bill with the coverage of that in Maine and with higher deposits than those now used in bottle bill states would increase recycling greatly and provide more jobs at the same time.

The European Commission has proposed a takeback law covering electronic equipment, including appliances, office machinery, and telecommunications equipment.

3.4 OVERALL PICTURE

For a sustainable future,[177] more things will have to be made from renewable resources and fewer from nonrenewable resources. This means using paper instead of plastic wherever possible, unless the plastic is based on a renewable source. The throwaway habit must be thrown away in favor of reusable objects, designed for long life, easy repair, and ease of recycling of the materials in them. Objects made of 100% postconsumer waste must become common, instead of being rare as they are today.

No one has found a way to curb the overconsumption of industrialized nations. (E. O. Wilson pointed out that if all nations used materials at the same rate as the United States, it would take the natural resources of two more Planet Earths.[178]) First, society must agree that this is a necessary objective. Next, some system of incentives to reduce and disincentives to use lavishly must be devised. This is the "carrot–stick" approach. This will involve a culture change.[179] Persons who live in the United States are constantly bombarded with messages to buy. The advertising to do this is US$500/person/year. Credit card companies deluge the average person with offers of cards that make buying on credit easy. The message is "Buy now–Pay later–Why wait to have what you want?" This makes the objects purchased cost significantly more. One can think of restrictions on advertising or taxes on advertising and credit cards, but these could be very unpopular. An approach that has increased the savings rate in some countries is similar to the individual retirement account (IRA) used in the United States. Money is set aside before taxes for the purchase of a first home or for a college education. There are no taxes until the money is withdrawn for the intended purpose. Premature withdrawal for another purpose imposes a tax penalty. (The individual savings rate in the United States is significantly lower than that in most other developed nations.)

In a finite world, there will be limits to what people can expect to use.[180] This means that the constant growth in the sales of a product, that some executives seem to expect, is unrealistic. One can see these unrealistic expectations expressed in almost any issue of *Chemical Engineering News*. The time may come when closed landfills are mined for the materials in them.

PROBLEMS

3.1 Cost-Effective Handling of Sewage Waste

You are in charge of waste (sewage) disposal for a metropolitan area populated by three million people. Several industrial firms are located in this area. You have found what you believe to be the most efficient and cost-effective approach for handling the problem of waste. The sewage is given primary treatment (i.e., lumps are filtered out) before discharge to a river that soon enters an estuary. The resulting sludge plus municipal solid waste is loaded onto barges for ultimate disposal in the ocean. Recently, you have been getting complaints from environmentalists who advocate more expensive waste disposal methods (which they do not have to pay for!). What should you do about this?

3.2 The Computer Age

Computers, cell phones, and other related gadgets are now ubiquitous. Progress in the technology is rapid, which leads to a high rate of obsolescence. The average personal computer contains the elements Hg, Pb, Ba, B, Co, Cu, as well as P or Br in its housing. Circuit boards may contain Cd and Au. Most adults in North America own a cell phone, although their average life is only 18 months. Each one may contain As, Sb, Be, Cd, Cu, Pb, Ni, Zn, and Br. Computers are replaced every three years, while they are still operable. People want the latest, largest flat-screen television set.

Putting obsolete electronic equipment in landfills is a waste of valuable materials, including ones which the world may run out of in 10–15 years. There is also the question of what may leach out from these products and potentially contaminate groundwater. Ones that are recycled end up in China or Nigeria, where the metals are recovered under very unsafe conditions. Devise some approaches to address this problem.

3.3 A Printing Challenge

You are the manager of a company called Print-All which, among other things, prints brightly colored gift packaging. The business is making a good profit and needs to expand. Since the process uses solvent-based inks, the expansion will add several tons to the solvent emissions over one year. To offset this, you have bought emission trading credits from a cleaner plant about 60 miles away in another city. (This is a legal process that has helped to reduce levels of emissions of sulfur dioxide from power plants.) However, the neighbors are not quite sure that they want the extra solvent dumped on them. Due to the prevailing winds and since the plant is on the edge of the city, most of the effects of the emissions will be in another location where the regulators have no control.

There are some ways to remove the effluent air from the plant, but they may be expensive when used with the large volumes of air needed to remove the solvent from the plant. There are also alternative methods of printing that do not require solvents. You are not very familiar with them, and heard a few years ago that they are ineffective using the types of equipment you have, and that they might also cost more. You are reluctant to gamble on new technologies that might cause the loss of your bonus as both of your children are attending expensive colleges. How can you appease the neighbors without making the plant noncompetitive and jeopardizing your own future?

3.4   It Is Turkey Time

Over Thanksgiving weekend, millions of Americans will be eating home-cooked turkey. You wish that they would see the health benefits in turkey and eat it almost every day, since you raise turkeys on a contract basis for Candue, Inc.

There are some problems with raising turkeys commercially. They keep producing waste (feces and urine on wood chips) that you have to dispose of onto neighboring fields. This has not been a big problem until recently when nearby coastal residents began to complain that nutrients from the turkey waste had leached into the inland bays and caused them to eutrophy. They say that this has discouraged tourism, causing them to lose money. Environmentalists strongly fear that the eutrophication could lead to a reduction in native biodiversity. Even if the problem is eliminated today, it will take 20 years for the ground to be free of the nutrients.

What can you do as a small producer when the profit that you make on raising turkeys is relatively small? Is it your problem or Candue's problem? What could be done that would convert the turkey waste to something that could be sold at a profit, or simply fed back to the turkeys?

3.5   Toys and Games from Trash

Children are inherently curious, and can learn a great deal by playing. What interests a child will often be different from what their parents expect. For example, some parents thought that their children would love a battery-powered elephant that walked and talked. However, they were surprised and dismayed when the children ignored the elephant and played with the box it came in instead.

Many dollars and much material go into the design and production of toys, especially toward the end of the year. Many toys have a short lifetime or the children outgrow them; then they fill the attic or become waste. It should be possible to circumvent this process and start with household waste to devise toys and games for children of various ages. This would cost nothing except time and ingenuity. Ideally, these could be used to teach science, technology, engineering, and mathematics (STEM subjects) while the children are having fun and not realizing that they were being taught. There should be a stimulus for the child to create something of his or her own in the game. Devise such a toy or game. If you finish in five minutes, then devise some others.

3.6 A Problem of Waste from University

You are the director of the Department of Public Works for a small college town. The university is an important part of the town's economy. The number of students is roughly equal to the number of townspeople. On the other hand, the university is the source of some of your problems. When the students move out at the end of the school year, a lot of trash is left. The Department of Public Works has to supply extra trash pick-up trucks. It also sets up a special fenced site for students to bring their trash and unwanted furniture to. You appreciate that the students are in a hurry to get to their summer jobs, and may not have room in the car for what they have accumulated during the school year. However, you still hate to see good, usable food, clothing, and furniture discarded along with the real waste. What can you do to salvage these materials for reuse and recycle?

The City Council voted not to approve money to buy reusable items from students to be sold back to incoming freshmen, citing concerns regarding where the items would be stored. Your chances of getting the university to be responsible for the trash of students who live off campus are slim. Could the university set up an online trading system for the usable items? What can you do?

3.7 The Three R's of the Environment

You are the commissioner of public works for a small city. The landfill is getting full and while everyone likes to see the trash truck once a week, no one wants a landfill next to where they live. Application of the three R's (Reuse, Reduce, and Recycle) could help. You could publicize them with a catchy jingle. Remember the one designed to sell more carbonated sugar water: "Twice as much for a nickel too. Pepsi is the drink for you." There could be a contest with the music teachers and their students as the judges or monetary prizes.

Energy, time, and money could be saved by putting the landfill next to where the trash originated. Global warming is here now and it needs to be addressed. There could be lovely communities such as Landfill Estates, Landfill Acres, Landfill Farms, and Village of Trash. Nice scents such as lemon, lavender, and chocolate could be released at the landfill. Getting everyone to put a dog-proof, raccoon-proof compost pile to compost food scraps, lawn clippings, and leaves would also help. In addition, better product designs could reduce the amount of trash greatly. Corrugated cartons are used to ship almost everything. While their recycling rate is high, it takes money and a factory to do it.

Design a container that can be used over and over again. It could nest or collapse to a flat structure. Germany uses a standard beer bottle, so that the bottle can be reused just by putting on a new label. Your new container could have a bar code on it so that any company could reuse it with the computer keeping track of it. Do you think any of the possibilities listed above could be made to work? Are modifications needed?

3.8 A Plastic Bottle Opportunity

You are the owner of a plant that makes plastic water bottles. The profits are good and the business seems to be recession-proof. Yet, there are some impending threats of the horizon. San Francisco has banned the sale of bottled water. Countries are taxing or outlawing plastic bags for groceries. Suppose this were to happen to single-use throwaway plastic bottles. What would you do: sell out or diversify into a new line of business? Recycling rates are low. Should you purchase a recycling plant and push curbside recycling? Some legislators are fed up, and others have suggested raising the deposit on each bottle and extending the deposit-refund system for the entire country. Customers like the "convenience" of the single-use bottle. Distributors do not like to handle and store returned bottles, nor do they want to put in bottle washing machines that would be required with refillable bottles. Rather than return to glass bottles, could a refillable plastic bottle be developed? Glass is heavy and it can break if the bottle is dropped.

You have heard of a process used to coat the inside of a plastic bottle so that it can be refilled. Should you license it from Mitsubishi? The FDA has approved it. Since the process cuts the oxygen through the bottle tenfold, it can be used for beer. This could be a gold mine. Beer sales are good, especially in college towns. Besides, glass beer bottles are banned at sporting events. Perhaps their use could be extended to wine bottles and containers for food. Your children are pushing you to do something about global warming. Refillable plastic containers would reduce the carbon footprint of the bottle business. You could be a leader in the business while keeping profits up and avoiding layoffs and pay cuts. This would keep employee morale and productivity up. Should you offer a discount on a refill to get a bottle back? Should you lease the bottle so that you have more control over what happens to it?

REFERENCES

1. (a) G.M. Levy, ed., *Packaging in the Environment*, Blackie Academic and Professional, London, 1993, 150; (b) G. Matos and L. Wagner, *Ann. Rev. Energy Environ.*, 1998, *23*, 107.
2. U.S. Environmental Protection Agency, Characterization of Municipal Solid Waste in the United States: 1994 Update. EPA 530-S-94-042, Nov. 1994 and 1996 update.
3. (a) A. Levey and E. Yermoli, *Waste Age's Recycling Times*, 1998, *10*(24), 3; (b) K. Egan, *Waste Age's Recycling Times*, 1998, *10*(11), 8.
4. K.M. White, *Waste Age's Recycling Times*, 1997, *9*(7), 7.
5. (a) S. Cassel, *Technol. Rev.*, 1992, *95*(6), 20; (b) ID2 Communications, Victoria, British Columbia, Canada gives 100–200 lb per worker per year, 2008.
6. P. Calmels and R. Harris, *Pulp Pap. Int.*, 1994, *36*(12), 47.
7. Anon., *R&D Cahners*, 2005, *47*(12), 32.
8. L. Zhito, www.americanprofile.com.
9. M. Chester, E. Martin, and N. Sathaye, *Environ. Sci. Technol.*, 2008, *42*, 2142.
10. J.M. Heumann, *Waste Age's Recycling Times*, 1998, *10*(7), 2.
11. D.J. Hanson, *Chem. Eng. News*, Feb. 26, 1996, 22.
12. K.A. O'Connell, *Waste Age's Recycling Times*, 1998, *10*(17), 6.
13. Delaware Economic Development Office, *Delaware Econom. Dev.*, Dover, DE, 1996, *1*(3), 8.
14. (a) J.T. Aquino, ed., *Waste Age/Recycling Times Recycling Handbook*, Lewis Publishers, Boca Raton, FL, 1995; (b) C. Boener and K. Chilton, *Kirk–Othmer Encyclopedia of Chemical Technology*, 4th ed., Wiley, New York, 1996, *20*, 1075; (c) A.G.R. Manser and A.A. Keeling, *Practical Handbook of Processing and Recycling Municipal Waste*. CRC Press, Boca Raton, FL, 1996; (d) C.P. Rader, ed., *Plastics, Rubber and Paper Recycling: A Pragmatic Approach*. ACS Symp. 609, Washington, DC, 1995; (e) A.K.M. Rainbow,

ed., *Why Recycle?* A.A. Balkema, Rotterdam, 1994; (f) C.R. Rhyner, L.J. Schwartz, R.B. Wenger, M.G. Kohrell, *Waste Management and Resource Recovery*, Lewis Publishers, Boca Raton, FL, 1995; (g) S.M. Turner, *Recycling*, American Chemical Society, Washington, DC, Dec. 1993; (h) R. Waite, *Household Waste Recycling*, Earthscan, London, 1995; (i) J. Carless, *Taking Out the Trash: A No-Nonsense Guide to Recycling*, Island Press, Washington, DC, 1992; (j) H.F. Lund, ed., *The McGraw-Hill Recycling Handbook*, 2nd ed., Hightstown, NJ, 2000.

15. J.C. Roberts, *The Chemistry of Paper*, Royal Society of Chemistry, Cambridge, 1996.

16. J.T. Staley and W. Hairpin, *Kirk–Othmer Encyclopedia of Chemical Technology*, 4th ed., Wiley, New York, 1992, *2*, 212.

17. M.S. Reisch, *Chem. Eng. News*, Dec. 10, 2007, 24.

18. www.innov-xsys.com.

19. G. Parkinson, An automated plant using similar techniques started in Hanover, Germany in 2000, *Chem. Eng.*, 1999, *106*(7), 19.

20. (a) F. Berman, *Trash to Cash*, St. Lucie Press, Delray Beach, FL, 1996; (b) C.G. Thompson, *Recycled Papers—The Essential Guide*, MIT, Press, Cambridge, MA, 1992.

21. B. Bateman, *Pap. Technol.*, 1996, *37*(1), 15.

22. (a) A. Karma, J. Engstrom, and T. Kutinlahti, *Pulp. Pap. Can.*, 1994, *95*(11), 38; (b) U. Arena, M.L. Mastellone, F. Perugini, and R. Clift, *Ind. Eng. Chem. Res.*, 2004, *43*, 5702.

23. B. Fransson and I. Emanuelsson, *Pap. Technol.*, 2002, *43*(7), 40.

24. M. Shaw, *Pulp and Paper*, Nov. 2002, 29.

25. (a) T. Pfitzner, *Pulp Pap. Int.*, 2003, *45*(4), 18; (b) J. Toland, *Pulp Pap. Int.*, 2003, *45*(4), 25.

26. T. Blain and J. Grant, *Pulp Can.*, 1994, *95*(5), 43.

27. (a) J.K. Borchard, *Kirk–Othmer Encyclopedia of Chemical Technology*, 4th ed., Wiley, New York, 1997, *21*, 10; *Chem. Ind. (Lond.)*, 1993, 273; *Prog. Pap. Recycl.*, 1994, *3*(4), 47; (b) L.D. Ferguson, *TAPPI J.*, 1992, *75*(7), 75; *75*(8), 49; (c) F.R. Hamilton, *PIMA Mag.*, Jan. 1992, 20; (d) T. Woodward, *PIMA Mag.*, Jan. 1992, 34; (e) A. Johnson, *Pap. Technol.*, 1992, *33*(11), 20; (f) C. Silverman, *Am. Ink Maker*, 1991, *69*(11), 28; (g) R.M. Rowell, T.L. Laufenberg, and J.K. Rowell, eds, *Materials Interactions Relevant to Recycling Wood-Based Materials*. Materials Research Society, Pittsburgh, PA, 1992; (h) F.J. Sutman, M.B.K. Letscher, and R.J. Dexter, *TAPPI J.*, 1996, *79*(3), 177; (i) R.C. Thompson, *Surf. Coat Int.*, 1998, *81*(5), 230; (j) M. Pescantin, *Pap. Technol.*, 2002, *43*(9), 47.

28. J.K. Borchardt, D.W. Matalamabi, V.G. Lott, and D.B. Grimes, *TAPPI J.*, 1997, *80*(10), 269.

29. S. Ritter, *Chem. Eng. News*, Nov. 16, 1998, 35.

30. P. Somasundaran, L. Zhang, S. Krishnakumar, and R. Slepetys, *Prog. Pap. Recycl.*, 1999, *8*(3), 22.

31. G. Sun and Y. Deng, *Prog. Pap. Recycl.*, 2000, *10*(1), 13.

32. M.S. Mahagaonkar, K.R. Stack, and P.W. Banham, *TAPPI J.*, 1998, *81*(12), 101.

33. J. Milliken, *TAPPI J.*, 1997, *80*(9), 79.

34. S. Dessureault, P. Carabin, A. Thom, J. Kleuser, and P. Gitzen, *Prog. Pap. Recycl.*, 1998, *8*(1), 23.

35. M.M. Sain, C. Daneualt, and M. Lapoint, *Am. Ink Maker*, 1996, *74*(4), 26.

36. G. Rangamannar and L. Silveri, *TAPPI J.*, 1990, *73*(7), 188.

37. (a) Anon., *Prog. Pap. Recycl.*, 1996, *6*(1), 73; (b) J. Bredael, N.J. Sell, and J.C. Norman, *Prog. Pap. Recycl.*, 1996, *6*(1), 24.

38. (a) J.D. Taylor and E.K.C. Yu, *Chemtech.*, 1995, *25*(2), 38; (b) A.K. Sharma, W.K. Forester, and E.H. Shriver, *TAPPI J.*, 1996, *79*(5), 211; (c) F. Ruzinsky and B.V. Kokta, *Prog. Pap. Recycl.*, 1998, *7*(3), 47; 2000, *9*(2), 30.

39. (a) J. Woodward, *Biotechnology*, 1994, *12*, 905; (b) R. Dinus and T. Welt, *Prog. Pap. Recycl.*, 1994, *3*(4), 63, 64.

40. (a) O.U. Heise, J.P. Unwin, J.H. Klungness, W.G. Fineran, Jr., M. Sykes, and S. Abubakr, *TAPPI J.*, 1996, *79*(3), 207; (b) P. Bajpai and P.K. Bajpai, *TAPPI J.*, 1998, *81*(12), 111; (c) A. Roring and R.D. Haynes, *Prog. Pap. Recycl.*, 1998, *7*(3), 73; (d) J.M. Jobbins and N.E. Franks, *TAPPI J.*, 1997, *80*(9), 73; (e) A.L. Morkbak and W. Zimmermann, *Prog. Pap. Recycl.*, 1998, *7*(3), 33; (f) S. Vyas and A. Lachke, *Enzyme Microb. Technol.*, 2003, *32*, 236.

41. G. Stork, H. Pereira, T.M. Wood, E.M. Dusterhoft, A. Toft, and J. Puls, *TAPPI J.*, 1995, *78*(2), 79, 89.

42. H. Pala, M.A. Lemos, M. Mota, and F.M. Gama, *Enzyme Microb. Technol.*, 2001, *29*, 274.

43. G. Parkinson, *Chem. Eng.*, 2002, *100*(11), 19.

44. N.W. Lazorisak, J.F. Schmitt, and R. Smith, U.S. Patent 5,620,565, 1997.

45. T. Friberg and G. Brelsford, *Solutions (from TAPPI and PIMA)*, Aug. 2002, 27.

46. K. Cathie, *Pap. Technol.*, 1997, *38*(5), 34.

47. Anon., *Pap. Technol.*, 1995, *36*(7), 13.

48. K.L. Patrick, *Pulp Pap.*, 1994, *68*(13), 75.

49. J. Schultz, *Paperboard Packag.*, 1995, *80*(2), 16.

50. C.L. Reynolds, *TAPPI J.*, 1997, *80*(9), 102.

51. L.R. Brown, *World Watch*, 1999, *12*(2), 13.

52. S. Westergard, *Pap. Technol.*, 1995, *36*(8), 28.

53. N. Wiseman and G. Ogden, *Pap. Technol.*, 1996, *37*(1), 31.

54. (a) D.G. Meadows, *TAPPI J.*, 1996, *79*(1), 63; (b) R.T. Klinker, *TAPPI J.*, 1996, *79*(1), 97.

55. J. Osborne, *Pulp Pap. Int.*, 1994, *36*(4), 100.

56. M.R. Servos, K.R. Munbittrick, J.H. Carey, and G.J. Van Der Krick, eds, *Environmental Fate and Effects of Pulp and Paper Mill Effluents*. St. Lucie Press, Delray Beach, FL, 1996.

57. H.-W. Liu, S.-N. Lo, and H.-C. Lavallee, *TAPPI J.*, 1996, *79*(5), 145.

58. (a) J. Brandrup, M. Bittner, W. Michaeli, and G. Meges, eds, *Recycling and Recovery of Plastics,* Hanser–Gardner, Cincinnati, OH, 1996; (b) R.J. Ehrig, *Plastics Recycling*, Hanser–Gardner, Cincinnati, OH, 1992; (c) A.L. Bisio and M.Xanthos, *How to Manage Plastics Waste*, Hanser–Gardner, Cincinnati, OH, 1994; (d) W. Heitz, ed., *Macromol. Symp.*, 1992, *57*, 1–395; (e) N. Basta, G. Ondrey, R. Rakagopal, and T. Kamiya, *Chem. Eng.*, 1997, *104*(6), 43; (f) League of Women Voters. *The Plastics Waste Primer: A Handbook for Citizens*, Washington, DC, 1993; (g) G.L. Nelson, *Chemtech*, 1995, *25*(12), 50; (h) J. Scheirs, *Polymer Recycling: Science, Technology and Applications*, Wiley, New York, 1998; (i) J. Pospisil, S. Nespurek, R. Pfaendner, and H. Zweifel, *Trends Polym. Sci.*, 1997, *5*, 294; (j) N.T. Dintcheva, N. Jilov, and F.P. LaMantia, *Polym. Degrad. Stabil.*, 1997, *57*, 191; (k) C. Crabb, *Chem. Eng.*, 2000, *107*(6), 41; (l) J. Brandrup, *Macromol. Symp.*, 1999, *144*, 439; (m) J. Kahovec, ed., *Macromol. Symp.*, 1998, *135*, 1–373; (n) A.L. Andrady, *Plastics and the Environment*, Wiley, New York, 2003; (o) A. Azapagic, A. Elmsley, and I. Hamerton, *Polymers: The Environmental and Sustainable Development*, Wiley, Chichester, 2003; (p) K.C. Frisch, D. Klempner, and G. Prentice, *Advances in Plastics Recycling*, 2 vols, CRC Press, Boca Raton, FL, 2001.

59. H. Ridgley, *Waste Age's Recycling Times*, 1997, *9*(17), 6.

60. (a) M.S. Reisch, *Chem. Eng. News*, May 22, 1995, 41; (b) K. Egan, *Waste Age's Recycling Times*, 1997, *9*(23), 1; (c) R.A. Denison, *Something to Hide: The Sorry State of Plastics Recycling*, *Environmental Defense Fund*, Washington, DC, Oct. 21, 1997; (d) *EDF Lett. (Environmental Defense Fund)*, 1998, *28*(1), 2.

61. (a) Anon., *Chem. Eng. News*, July 22, 1996, 26; (b) C. Rovelo, *Waste Age's Recycling Times*, 1997, *9*(8), 15; (c) Anon., *Chem. Eng. News*, Dec. 9, 1996, 18; (d) E.M. Kirschner, *Chem. Eng. News*, Nov. 4, 1996, 19; (e) Anon., *Chem. Eng. News*, June 23, 1997, 15; (f) J.M. Heumann, *Waste Age's Recycling Times*, 1997, *9*(13), 14; (g) J.L. Bast, *Waste Age's Recycling Times*, 1997, *9*(13), 15; (h) D.L. NaQuin, *Waste Age's Recycling Times*, 1997, *9*(18), 13; (i) Anon., *Chem. Eng. News*, Sept. 28, 1998, 14; (i) Anon., *Chem. Eng. News*, Sept. 3, 2001, 15.

62. (a) W. Weizer, *Chem. Ind. (Lond.)*, 1995, 1013; (b) R.A. Denison, *EDF Lett. (Environmental Defense Fund)*, New York, 1993, *24*(6), 4; (c) H. Hansler, *Chem. Ind. (Lond.)*, 2000, 429.

63. (a) G.P. Karayannidis, D.E. Kokkalas, and D.N. Biliaris, *J. Appl. Polym. Sci.*, 1995, *56*, 405; (b) A.H. Tullo, *Chem. Eng. News*, Oct. 15, 2007, 15; (c) H. Hansler, *Chem. Ind. (Lond.)*, July 10, 2000, 429.

64. D. O'Sullivan, *Chem. Eng. News*, Feb. 12, 1990, 25.

65. (a) J.M. Heumann, *Waste Age's Recycling Times*, 1998, *10*(17), 23; (b), M. McCoy, *Chem. Eng. News*, Mar. 16, 2009, 30.

66. Anon., *American Chemistry*, Mar./Apr. 2007, 26.

67. (a) K. Egan, *Waste Age's Recycling Times*, 1998, *10*(9), 1; (b) M. Paci and F.P. LaMantia, *Polym. Degard. Stab.*, 1999, *63*, 11.

68. (a) J.M. Heumann, *Waste Age's Recycling Times*, 1998, 10(9), 1; (b) S. Ulutan, *J. Appl. Polym. Sci.*, 1998, *69*, 865; (c) D. Braun, *Prog. Polym. Sci.*, 2002, *27*, 2171, 2183.

69. (a) Anon., *Chem. Eng. News*, Mar. 4, 2002, 15; Jan. 19, 2004, 17; (b) A. Scott, *Chem. Week*, Mar. 6, 2002, 31.

70. (a) S.E. Selke, *Packaging and the Environment: Alternatives, Trends and Solutions*, 2nd ed., Technomic, Lancaster, PA, 1994; (b) A. Brody and K. Marsh, *The Wiley Encyclopedia of Packaging Technology*, 2nd ed., Wiley, New York, 1997; (c) Anon., *Food Technol.*, 1990, *44*(7), 98; (d) K. Dotson, *Int. News Fats Oils Relat. Mater*, 1991, *2*, 854.

71. J.K.S. Wan, K.P. Vepsalainen, and M.S. Ioffe, *J. Appl. Polym. Sci.*, 1994, *54*, 25.

72. (a) Anon., *R&D (Cahners)*, 1998, *40*(10), 130; (b) M. Jacoby, *Chem. Eng. News*, Aug. 16, 1999, 38.

73. Anon., *Chem. Eng. News*, Feb. 13, 1995, 11.

74. H. Ridgley, *Waste Age's Recycling Times*, 1997, *9*(19).

75. D. Staicu, G. Banica, and S. Stoica, *Polym. Degrad. Stabil.*, 1994, *46*, 259.

76. (a) R. Pfaendner, H. Herbst, K. Koffmann, and F. Sitek, *Angew. Makromol. Chem.*, 1995, *232*, 193; (b) M.K. Loultchea, M. Proietto, N. Jilov, and F.P. la Mantia, *Polym. Degrad. Stabil.*, 1997, *57*, 77; *Macromol. Symp.*, 2000, *152*, 201; (c) J. Pospisil, S. Nespurek, R. Pfaenner, and H. Zweifel, *Trends Polym. Sci.*, 1997, *5*, 294; (d) F.P. LaMantia, *Macromol. Symp.*, 2000, *152*, 201; (e) C.N. Kartallis, C.D. Papaspyrides, and R. Pfaendner, *J. Appl. Polym. Sci.*, 2003, *88*, 3033.

77. L. Willis, *Chemtech.*, 1994, *24*(2), 51.

78. H. Ridgley, *Waste Age's Recycling Times*, 1997, *9*(20), 7.

79. M. Xanthos, B.C. Baltzis, and P.P. Hsu, *J. Appl. Polym. Sci.*, 1997, *64*, 1423.

80. Anon., *Chem. Br.*, 1996, *32*(3), 11.

81. M. Paci and F.P. la Mantia, *Polym. Degrad. Stabil.*, 1998, *61*, 417.

82. S. Dutta and D. Lohse, *Polymeric Compatibilizers: Uses and Benefits in Polymer Blends*, Hanser–Gardner, Cincinnati, OH, 1996.

83. (a) J. Duvall, C. Selliti, V. Topolkaraev, A. Hiltner, E. Baer, and C. Myers, *Polymer*, 1994, *35*, 3948; (b) G.H. Kim, S.S. Hwang, B.G. Cho, and S.M. Hong, *Macromol. Symp.*, 2007, *249–250*, 485.

84. N.K. Kalfoglou, D.S. Skafidas, and D.D. Sotiropoulou, *Polymer*, 1994, *35*, 3624.

85. P. van Ballegovic and A. Rudin, *J. Appl. Polym. Sci.*, 1990, *39*, 2097.

86. T.J. Nosker, D.R. Morrow, R.W. Renfree, K.E. Van Ness, and J.J. Donaghy, *Nature*, 1991, *350*, 563.

87. (a) E. Kroeze, G. ten Brinke, and G. Hadziioannou, *Polym. Bull.*, 1997, *38*, 203; (b) S.C. Tjong and S.A. Xu, *J. Appl. Polym.Sci.*, 1998, *68*, 1099.

88. G. Radonji and C. Musil, *Angew. Makromol.Chem.*, 1997, *251*, 141.

89. W. Hoyle and D.R. Karsa, *Chemical Aspects of Plastic Recycling*, Special Publication 199, Royal Society of Chemistry, Cambridge, 1997.

90. (a) G.K. Wallace, *TAPPI J.*, 1996, *79*(3), 215; (b) D. Paszun and T. Spychaj, *Ind. Eng. Chem. Res.*, 1997, *36*, 1373; (c) D.-C. Wang, L.-W. Chem, and W.-Y. Chiu, *Angew. Makromol. Chem.*, 1995, *230*, 47; (d) G. Parkinson, *Chem. Eng.*, 1996, *103*(3), 19; (e) G. Samadani, *Chem. Eng.*, 1995, *102*(5), 15; (f) F. Johnson, D.L. Sikkenga, K. Danawala, and B.I. Rosen, U.S. Patent 5,473, 102, 1995; (g) G. Guclu, A. Kasgoz, S. Ozbudak, S. Ozgumus, and M. Orbay, *J. App. Polym. Sci.*, 1998, *69*, 2311; (h) G. Parkinson, *Chem. Eng.*, 1999, *106*(12), 21; (i) J.-W. Chen and L.-W. Chen, *J. Appl. Polym. Sci.*, 199, *73*, 35; (j) B.-K. Kim, G.-C. Hwang, S.-Y. Bai, S.C. Yi, and H. Kumazawa, *J. Appl. Polym. Sci.*, 2001, *81*, 2102; (k) G.P. Karayannidis and D.S. Axilias, *Macromol. Mater. Eng.*, 2007, *292*, 128–146.

91. A. Krazn, *J. App. Polym. Sci.*, 1998, *69*, 1115.

92. H. Wang, L. Chen, X. Liu, Y. Zhang, Z. Wu, and Y. Zhou, *Polym. Preprints*, 2002, *43*(1), 473.

93. K.A. O'Connell, *Waste Age's Recycling Times*, 1997, *9*(19), 15.

94. H. Winter, H.A.M. Mostert, P.J.H.M. Smeets, and G. Paas, *J. Appl. Polym. Sci.*, 1995, *57*, 1409.

95. K.H. Yoon, A.T. DiBenedetto, and S.J. Huang, *Polymer*, 1997, *38*, 2281.

96. S.J. Shafer, U.S. Patent 5,336,814, 1993.

97. R. Pinero, J. Garcia, and M.J. Cocero, *Green Chem.*, 2005, *7*, 380.

98. G. Ondrey, *Chem. Eng.*, 2004, *111*(13), 14.

99. H. Mueller, *Chem. Abstr.*, 1996, *124*, 9639.

100. (a) G. Scott and D. Gilead, eds, *Degradable Polymers: Principles and Applications*, Chapman & Hall, New York, 1996; (b) G.J.L. Griffin, *Chemistry and Technology of Biodegradable Polymers*, Chapman & Hall, London, 1994; (c) S.J. Huang, ed., *Polym. Degrad. Stabil.*, 1994, *45*(2), 165–249; (d) A.-C. Albertsson and S. Karlsson, eds, *Macromol. Symp.*, 1998, *130*, 1–410; (e) J. Kahovec, ed., *Macromol. Symp.*, 1997, *123*, 1–249; (f) R. Chandra and R. Rustgi, *Prog. Polym. Sci.*, 1998, *23*, 1273; (g) R.A. Gross and B. Kalra, *Science*, 2002, *297*, 803; (h) E.S. Stevens, *Green Plastics*, Princeton University Press, Princeton, 2002.

101. W. Rathje and C. Murphy, *Rubbish: The Archaeology of Garbage*, Harper–Collins, New York, 1992.

102. (a) R. Arnaud, P. Dabin, J. Lemaire, S. Al-Malaika, S. Chohan, M. Coker, G. Scott, A. Fauve, and A. Maaroufi, *Polym. Degrad. Stabil.*, 1994, *46*, 211; (b) B.S. Yoon, M.H. Suh, S.H. Cheong, J.E. Yie, S.H. Yoon, and S.H. Lee, *J. Appl. Polym. Sci.*, 1996, *60*, 1677; (c) B.G. Kang, S.H. Yoon, S.H. Lee, J.E. Yie, B.S. Yoon, and M.H. Shu, *J. Appl. Polym. Sci.*, 1996, *60*, 1977.

103. Anon., *Waste Age's Recycling Times*, 1998, *10*(7), 5.

104. J. Trittin, *Environ. Sci. Technol.*, 1999, *33*, 194A.

105. (a) H.V. Makar, *Kirk–Othmer Encyclopedia of Chemical Technology*, 4th ed., Wiley, New York, 1996, *20*, 1092; (b) W.H. Richardson, *Kirk–Othmer Encyclopedia of Chemical Technology*, 4th ed., Wiley, New York, 1996, *20*, 1106; (c) I.K. Wernick and N.J. Themeils, *Ann. Rev. Energy Environ.*, 1998, *23*, 465; (d) S.K. Ritter, *Chem. Eng. News*, June 8, 2009, 53.

106. A.L. Brody, *Kirk–Othmer Encyclopedia of Chemical Technology*, 4th ed., Wiley, New York, 1994, *11*, 834.

107. K. Egan, *Waste Age's Recycling Times*, 1997, *9*(8), 7.
108. K.M. White, *Waste Age's Recycling Times*, 1998, *10*(8), 1.
109. Anon., Business Wire, July 21, 2008.
110. Anon., *World Watch*, 2004, *17*(1), back cover.
111. B.J. Jody, E.J. Daniels, and N.F. Brockmeier, *Chemtech.*, 1994, *24*(11), 41.
112. Anon., *Waste Age's Recycling Times*, 1998, *10*(24), 5.
113. T.J. Veasey, R.J. Wilson, and D.M. Squires, *The Physical Separation and Recovery of Metals from Wastes*, Gordon & Breach, Newark, NJ, 1993.
114. (a) K.C. Frisch, D. Klempner, and G. Prentice, *Recycling of Polyurethanes*, CRC Press, Boca Raton, FL, 1999; (b) W. Rabhofer and E. Weigand, *Automotive Polyurethanes*, CRC Press, Boca Raton, FL, 2001.
115. M. Freemantle, *Chem. Eng. News*, Nov. 27, 1995, 25.
116. G. Ondrey, *Chem. Eng.*, 2000, *107*(11), 21.
117. G.S. Samdani, *Chem. Eng.*, 1995, *102*(6), 17.
118. G. Parkinson, *Chem. Eng.*, 1998, *105*(11), 23.
119. K. Murase, K.-I. Nishikawa, K.-I. Machida, and G.-Y. Adachi, *Chem. Lett.*, 1994, *23*, 1845.
120. S. Akita, T. Maeda, and H. Takeuchi, *J. Chem. Technol. Biotechnol.*, 1995, *62*, 345.
121. (a) C.P. Ross, *Kirk–Othmer Encyclopedia of Chemical Technology*, 4th ed., Wiley, New York, 1996, *20*, 1127; (b) M. Gruver, *Wilmington Delaware News J.*, Oct. 3, 2009, A8.
122. J.T.B. Tripp and A.P. Cooley, *EDF Lett. (Environmental Defense Fund)*, New York, 1996, *27*(6), 4.
123. R. Simon, *Forbes*, May 28, 1990, 136.
124. K.A. O'Connell, *Waste Age's Recycling Times*, 1998, *10*(14), 13.
125. R.M. Kashmanian and R.F. Rynk, *Am. J. Alt. Agric.*, 1998, *13*, 40.
126. P. Drechsel and D. Kunze, eds, *Waste Composting for Urban and Peri-Urban Agriculture*, Oxford University Press, Oxford, 2001.
127. D.J. Sessa and Y.V. Wu, *Int. News Fats Oils Relat. Mater.*, 1996, *7*(3), 274.
128. M.A. Martin-Cabrejas, R.M. Esteban, F.J. Lopez-Andreu, K. Waldron, and R.P. Selvendran, *J. Agric. Food Chem.*, 1995, *43*, 662.
129. Anon., *Environment*, 1994, *36*(10), 22.
130. (a) Anon., *Environ. Sci. Technol.*, 2003, *37*, 389A; 2004, *38*, 265A; (b) T. Staedler, *Technol. Rev.*, 2003, *106*(5), 73.
131. Information from Brookrock Corp. Newark, DE.
132. (a) S. Ottwell, *Chem. Eng. (Rugy, Engl.)*, 1996, (Feb.), 15; (b) C.J. Nagel, C.A. Chanechuk, E.W. Wong, and R.D. Bach, *Environ. Sci. Technol.*, 1996, *30*, 2155; (c) E. Rafferty, *Chem. Eng.*, 1997, *104*(11), 43; (d) J.J. Cudahy, *Environ. Prog.*, 1999, *18*, 285; (e) Molten Metal Technologies, Waltham, MA.
133. (a) D. NaQuin, *Waste Age's Recycling Times*, 1998, *10*(14), 14; (b) K. Bullis, *Technol. Rev.*, Mar./Apr. 2007, 25; (c) C.G. Daye, *Chicago AC Meeting*, Aug. 2001, SCHB 4; (d) D. Talbot, *Technol. Rev.*, 2004, *107*(1), 24; (e) P. McKenna, *New Sci.*, April 25, 2009, 33.
134. "Plasma Arc Technology," www.google.com, 2008.
135. Anon., *Chem. Eng. Prog.*, 2004, *100*(1), 15.
136. G. Parkinson, *Chem. Eng. Prog.*, 2005, *101*(7), 13.
137. (a) S. Tyson and T. Blackstock, *Paper Pap.-Am. Chem. Soc., Div. Fuel Chem.*, 1996, *41*(2), 587; C.A., 1996, *124*, 209, 949; (b) R.E. Hughes, G.S. Dreher, M. Rostam-Abadi, D.M. Moore, and P.J. DeMaris, *Paper. Pap.-Am. Chem. Soc., Div. Fuel Chem.*, 1996, *41*(2), 597; C.A., 1996, *124*, 239, 889.
138. D. NaQuin, *Waste Age's Recycling Time*, 1997, *9*(26), 14.
139. J. Beretka, R. Cioffi, L. Santoro, and G.L. Valenti, *J. Chem. Technol. Biotechnol.*, 1994, *59*, 243.
140. G. Parkinson, *Chem. Eng.*, 1997, *104*(5), 29.

141. (a) W. Zschau, *Int. News Fats Oils Relat. Mater.*, 1994, *5*(12), 1375; (b) C. Chung and V. Eidman, *Int. News Fats Oils Relat. Mater.*, 1997, *8*, 739.

142. J.H. Krieger and M. Fremantle, *Chem. Eng. News*, July 7, 1997, 10.

143. S. Strasser, *Waste and Want: A Social History of Trash*, Metropolitan Books, New York, 1999.

144. H. Petroski, *The Pencil—A History of Design and Circumstance*, Knopf, New York, 1990.

145. R.L. Maine, *Prog. Pap. Recycl.*, 1993, *2*(3), 76.

146. (a) S.E.M. Selke, *Packaging and the Environment*. Technomic, Lancaster, PA, 1990; 2nd ed., 1994; (b) E.J. Stilwell, R.C. Canty, P.W. Kopf, and A.M. Montrone, *Packaging for the Environment—A Partnership for Progress*, American Management Association, New York, 1991; (c) G.M. Levy, ed., *Packaging in the Environment*, Blackie Academic and Professional, London, 1993.

147. D. Saphire, *Benefits of Reusable Shipping Containers—Delivering the Goods*, Inform New York, 1995.

148. The Presidential Green Chemistry Challenge Awards Program, U.S. Environmental Protection Agency, Washington, DC, EPA 744-K-96-001, 1996, July 33.

149. Anon., *Pollout. Equip. News*, 1997, *30*(5), 4.

150. K.A. O'Connell, *Waste Age's Recycling Times*, 1997, *9*(24), 13.

151. K. Holler, *Waste Age's Recycling Times*, 1997, *9*(24), 10.

152. H. Ridgley, *Waste Age's Recycling Times*, 1997, *9*(23), 4.

153. Anon., *Environment*, 1997, *39*(3), 23.

154. (a) Letter. G. T. Kimbrough of Ashland Superamerica to A.S. Matlack, 2/15/96; (b) *Packaging*, 1995, *66*(705), 13.

155. K.M. Reese, *Chem. Eng. News*, June 8, 1998, 64.

156. H.E. Allen, M.A. Henderson, and C.N. Haas, *Chemtech.*, 1991, *21*(12), 738.

157. Anon., *Amicus J.* (Natural Resources Defense Council), 2000, *22*(2), 7.

158. Anon., *Tufts Univ. Health Nutr. Lett.*, 1997, *15*(7), 3.

159. Anon., *Package Dig.*, 1995, *32*(2), 68.

160. (a) E. Royte, *Bottlemania: How Water Went on Sale and We Bought It*, Bloomsburg, New York, 2008; (b) J. Motavalli, *EDF Solutions*, 2008, *39*(4) 10; (c) E. Arnold and J. Larsen, *Earth Policy Institute*, Feb. 2, 2006; (d) Anon., *Tufts Univ. Health Nutr. Lett.*, 2007, *25*(3), 7.

161. M. Smith, *The U.S. Paper Industry and Sustainable Production*. MIT Press, Cambridge, MA, 1997.

162. J. Williams, *Wilmington Delaware News J.*, Apr. 25, 1996, D1.

163. R. Graff and B. Fishbein, *Reducing Office Paper Waste*, Inform, New York, 1991.

164. (a) G.L. Nelson, *Chemtech.*, 1995, *25*(12), 50; (b) R.A. Denison, *Ann. Rev. Energy Environ.*, 1996, *21*, 191.

165. W. McDonough and M. Braungart, *Cradle to Cradle: Remaking the Way That We Make Things*, North Point Press, New York, 2002.

166. M.B. Hocking, *Science*, 1991, *251*, 504.

167. Anon., *Packag Dig.*, 1995, Nov. 42.

168. F. Ackerman, *Why Do We Recycle? Markets, Values and Public Policy*. Island Press, Washington, DC, 1997.

169. M.S. Andersen, *Environment*, 1998, *40*(4), 10.

170. K.A. O'Connell, *Waste Age's Recycling Times*, 1997, *9*(5), 1; *9*(6), 7.

171. F. Ackerman, *Environment*, 1992, *34*(5), 2.

172. R.F. Stone, A.D. Sagar, and N.A. Ashford, *Technol. Rev.*, 1992, *95*(5), 48.

173. S. Odendahl, *Pulp Pap. Can.*, 1994, *95*(4), 30.

174. *Environ. Sci. Technol.*, 1995, *29*(9), 407A.

175. D. Saphire, *Case Reopened—Reassessing Refillable Bottles*, Inform, New York, 1994, 222, 233, 244, 264.

176. K.M. White, *Waste Age's Recycling Times*, 1997, *9*(10), 6.

177. (a) R.W. Kates, *Environment*, 2000, *42*(3), 10; (b) P.F. Barlett and G.W. Chase, eds, *Sustainability on Campus—Stories and Strategies for Change*, Island Press, Washington, DC, 2004; (c) T.E. Graedel and R.J. Klee, *Environ. Sci. Technol.*, 2002, *36*, 523; (d) S. Dresner, *The Principles of Sustainability*, Stylus Publishing, Herndon, VA, 2002.

178. E.O. Wilson, *Science*, 1998, *279*, 2048.

179. (a) G. Gardner and P. Sampat, *Mind Over Matter: Recasting the Role of Materials in Our Lives*. Worldwatch Paper, 144, World Watch Institute, Washington, DC, 1998; (b) J.B. Schor, *The Overspent American—Why We Want What We Don't Need*, Harper Perennial, New York, 1999.

180. (a) D.H. Meadows, D.L. Meadows, and J. Randers, *Beyond the Limits*, Chelsea Green Publishers, Post Mills, VT, 1992; (b) A. Durning, *How Much Is Enough? The Consumer Society and the Future of the Earth*, Norton, New York, 1992; (c) R.W. Kates, *Environment*, 2000, *42*(3), 10; (d) T. Jackson, *The Earthscan Reader on Sustainable Consumption*, Stylus Publishing, Herndon, VA, 2006.

4 Energy and the Environment

4.1 ENERGY-RELATED PROBLEMS[1]

Energy is needed for heating, cooling, and lighting homes, offices, and manufacturing plants; cooking; transportation; farming; manufacturing of goods; and a variety of other uses. Buildings in the United States use 36% of the total energy and 65% of the electricity, produce 30% of the greenhouse emissions, use 30% of the raw materials, produce 30% of the waste and consume 12% of the potable water.[2] Homes in the United States use 46% of their energy for space heating, 15% for heating water, 10% for food storage, 9% for space cooling, 7% for lighting, and 13% for other uses.[3] The corresponding figures for commercial buildings are 31%, 4%, 5%, 16%, 28%, and 16%. Transportation accounts for one-fourth of the energy consumption and two-thirds of the oil used in the United States. Industry uses 37% of the energy in the United States. Agriculture, including the transportation and processing of foods, uses about 17% of the energy in the United States. Industrialized societies tend to be energy intensive. Most of this energy is derived from fossil fuels. In 1993, oil provided 39% of the world's energy, natural gas 22%, and coal 26%, whereas hydropower and other renewable energy sources provided only 8%.[4] Nuclear power accounted for 5% of the total. In 2004, the world obtained 39% from oil, 24% from coal, 23% from natural gas, 7% from nuclear sources, and 7% from renewable sources of which 90% was hydropower.[5] This energy consumption has produced various problems, some of which are quite serious. Even nuclear energy and renewable energy have problems that need to be solved. Although technology can solve some of these problems, others may require changes in our lifestyles, habits, value systems, and outlooks. This chapter will focus primarily on the role that chemistry can play in solving some of the problems.[6] It will also mention areas in which social factors play important roles.

4.1.1 GENERATION OF ELECTRIC POWER[7]

The burning of coal is used to generate much of the electric power in the United States. For example, Delmarva Power obtains 55.9% of its power from coal, 7.3% from gas, 34.2% from nuclear sources, 0.5% from oil, and 2.1% from renewable sources.[8] Strip mining of coal turns the earth upside down. Although reclamation is now required, the original, diverse natural system is not replaced. Instead, a more restricted selection of plants is used. The soil may be quite acidic owing to oxidation of iron sulfide pyrite brought to the surface during the mining. The same process can result in acid mine drainage into local streams, resulting in a drastic reduction in their biota. Tailings ponds must be constructed so that no fine coal enters the stream and the dam is strong enough to withstand heavy rains and possible floods. Mining underground coal is the most hazardous occupation in the United States. Miners may be exposed to roof falls, dust explosions, methane explosions, black lung disease from breathing coal dust, and other problems.

Burning coal in the power plant produces sulfur dioxide, nitrogen oxides, fine particulate matter, and emissions of toxic heavy metals, such as mercury. Electric power generators are responsible for 78% of the sulfur dioxide, 64% of the nitrogen oxides, 40% of the fine particulates, 21% of the airborne mercury, and 35% of the carbon dioxide released in the United States.[9] Many old plants in existence when the Clean Air Act was passed were exempted from the regulations. It has been estimated that if the 559 dirtiest plants met the current source emissions standards, it would cut the emission of nitrogen oxides by 69% and that of sulfur dioxide by 77%.[10] Although emissions of sulfur dioxide, carbon monoxide, and lead have fallen since 1988, ground-level ozone has remained about the same, and nitrogen oxides are increasing.[11] Annual crop damage from ground-level ozone in the United States is about US$3 billion.[12]

Nitrogen oxides are produced when fuels are burned at high temperatures. The amount formed is less at lower temperatures. Catalytic combustion using a palladium catalyst has been studied as a way of burning the fuel at a lower temperature.[13] Nitrogen oxides can also be removed by reduction of the stack gases with ammonia, hydrogen, carbon monoxide, or alkanes using various catalysts to form nitrogen.[14] Nitrogen oxides formed during combustion react with hydrocarbons in the presence of light to form ground-level ozone. The U.S. EPA has proposed to require deep cuts (up to 85%) in the amount of nitrogen oxides emitted by power plants in the eastern United States as a way of reducing ground-level ozone.[15] This may require more than just burning at a lower temperature. The Edison Electric Institute, a trade association of utilities, calls the proposal "unfair, expensive and misdirected." The American Chemistry Council (formerly the Chemical Manufacturers Association) is also concerned about the proposal. Motor vehicles produce about half of the nitrogen oxides in the air. Power plants produce much of the rest. Ships at sea account for 14% of the total emissions of nitrogen oxides, from the burning of fossil fuels, and 16% of the sulfur dioxide arising from the burning of petroleum. Their residual oil fuel may contain as much as 4.5% sulfur.[16] Some of these emissions drift over and fall out on land in places such as Scandinavia. Acid rain from sulfur dioxide and nitrogen oxides emissions[17] has wiped out the fish populations in 9500 lakes in Norway with another 5300 remaining at risk.[18] Much of the emissions arise in other countries, such as the United Kingdom, and drift in with the winds.[19] (A study by the British government concluded that up to 24,000 persons die prematurely in the United Kingdom each year following short-term air pollution episodes involving ozone, particulates, and sulfur dioxide).[20] Simply reducing the supply of acid rain now may not be enough to restore these lakes, because the buffer capacity of the soil of the watershed may have been exhausted. Such depletion of calcium ions in the soil has been documented in the mountains of New Hampshire in the United States.[21] Acid rain also causes reductions in crop and timber yields, as well as corrosion of metals in cars, power lines, and the like, and of buildings made of limestone, marble, and sandstone. It is also partially responsible for the eutrophication of water in places such as the Chesapeake Bay in Maryland.

4.1.2 TRANSPORTATION

The 5% of the world's population living in the United States consumes 25% of its oil.[22] (The United States has only 3% of the world's oil reserves.[23]) The transportation sector uses 65% of this oil. Almost all (97%) transportation in the United States relies on oil. Personal automobiles consume 50% of the oil.[24] Transportation accounted for one-third of the carbon dioxide emissions of the United States in 2003.[25] Gasoline consumption was almost 114 billion gallons in 1996.[26] America's drivers burned an average of 336 million gallons of gasoline per day for the first 8 months of 1997.[27] Half of this oil was imported. In June 1992, the cost of this oil plus the importation of cars in which it could be used was US$6.6 billion.[28] Partly as a result of this, the United States is now the world's largest debtor nation.[29] Europe imports 76% of its petroleum.[30] Transportation produces 20% of all carbon dioxide in the world and 67% of the pollutant emissions in the world's cities.

Numerous environmental problems are associated with the production and use of this oil. Extensive studies have been made on the fate of the oil spilled in Alaskan waters from the *Exxon Valdez*.[31] Leaks in pipelines have also been problems.[32] Twenty-five years after a spill in Alberta, Canada, large amounts of oil were still present at depths of 10–40 cm. Oil can get into the water from the exhausts of older outboard motors, which burn a mixture of 1 quart of oil to 1 gallon of gasoline. There is also the problem of what to do with used drilling muds and oil well brines, especially on offshore drilling platforms. The methyl *tert*-butyl ether added to gasoline to reduce air pollution has shown up in shallow wells and springs in several urban areas.[33] This may have resulted both from spills and by evaporation or leakage from tanks. Most of it in Donner Lake, California, came from recreational boating.[34] The additive is not used any longer.[35] Ethanol is used instead.

Urban sprawl is a big part of the problem in the United States. Americans in cities drive five times as far as persons who live in European cities.[36] The United States has more cars per capita than any other developed nation.[37] An Australian has described the situation as "totally out of control." Only 3% of travel in the U.S. cities is by public transport. One-fourth of the travel in European cities is by this method, whereas it is two-thirds in Hong Kong and Tokyo. Close to half of all urban space in the United States is for cars. The average time to commute to work in the United States is 22 minutes.[38] Only 4% travel to work by public transportation. Only 3% walk to work.[39] Only 22% of car miles are to work, whereas about a third are to pick up children and to go shopping.[40] Many children are effectively immobilized by unsafe roads and the lack of sidewalks in suburbia, which prevent them from walking. Children old enough to drive cars often go to shopping malls to find a place to walk and meet their friends. Some of them also have a curious energy-intensive pastime of cruising their cars round and round a loop in the center of Newark, Delaware, on Friday night, as they watch their friends in other cars. Widening roads accounts for 44% of highway costs. Subsidies to roads are seven times those for public transportation. The difference between work and welfare in the inner city may exist, in part, because of inadequate public transportation.

Alternatives to sprawl include clustered housing and public transit–oriented development.[41] A return to more compact, village-like living has been proposed.[42] People in compact neighborhoods in San Francisco made 42% fewer trips by car than suburbanites.[43] Mixed-use developments cut auto use. Portland, Oregon, has urban growth boundaries and other laws that are leading to a more livable city.[44] Curitiba, Brazil, has an integrated public transportation system that functions well. The combination of cheap gasoline, few safe places to walk in suburbia, and abundant food has made one-third of Americans obese (i.e., 20% or more over optimum weight). World oil production was predicted to peak in 2010[45] (U.S. oil production peaked in 1970). This should lead to higher gasoline prices. To moderate the problems associated with global warming, it would be desirable to change to a less energy-intensive system before 2010.

4.1.3 Global Warming[46]

The world is now 0.8°C warmer than it was in the nineteenth century, according to the United Nations Intergovernmental Panel on Climate Change.[47] The year 1998 was the warmest year up until that time since record keeping began in 1856. It was subsequently surpassed by 2006 as the hottest year.[48] Based on tree rings, pollen, sediments, gases trapped in glaciers, and corals, it was the hottest year in the last 1000 years.[49] This is the greatest rate of temperature change in the last 10,000 years. Global minimum temperatures are rising faster than the maxima.[50] Temperatures in polar regions have risen faster than the global average (e.g., as much as 6°F in Alaska).[51]

Although the presence of aerosols from volcanoes and air pollution have complicated the study,[52] it is now believed to be due to human activities. It is due to putting more greenhouse gases, such as carbon dioxide, nitrous oxide, chlorofluorocarbons, hydrofluorocarbons, perfluorocarbons, and sulfur hexafluoride, into the atmosphere.[53] Sulfur hexafluoride is the most potent, being 23,900 times as potent as carbon dioxide.[54] Efforts are being made to reduce its use by the electrical and other industries, or at least to use systems to capture it for recycling. Carbon dioxide is by far the largest, amounting to 23 billion tons each year.[55] The preindustrial level of 280 ppm in the air has risen dramatically largely from the burning of fossil fuels.[56] This value could reach 500 ppm by the year 2100 and result in a rise of global temperature of 1.0°C–3.5°C. This warming was first predicted by Arrhenius 100 years ago.[57]

Along with the increase in temperature, by 2100 there will be a rise in sea level of 15–95 cm.[58] Glaciers and islands will shrink. More frequent extreme weather events such as droughts and floods are expected. A 1-m rise in sea level would threaten half of Japan's industrial areas. An estimated 118 million people may be at risk from the rising sea level.[59] Many species of plants and animals will have to migrate poleward, a process that may be difficult, considering the extensive habitat fragmentation in many countries. The responses to global warming can be seen in some species already. A gradual warming of the California current by 1.2°C–1.6°C since the 1950s has resulted in an 80% decline in the zooplankton, a 60% decline in Cassin's auklet, and a 90% decline in the sooty shearwater.[60] More winters without sea ice in Antarctica have caused a drop in the krill population.[61] Birds are laying their eggs earlier in the season in the United Kingdom.[62] The bleaching of corals on tropical reefs is considered by some to be an early warning sign of global warming.[63]

The range of Edith's checkerspot butterfly in the western United States, Mexico, and Canada is moving northward.[64] Absorption of carbon dioxide by the oceans is lowering the pH to a point where it may interfere with the calcification of animals.[65] There is also the fear that global warming could shift ocean currents, with the result that Europe could become much colder.[66]

Various suggestions have been made for sequestering some of the carbon dioxide. Planting trees (e.g., on degraded land) could remove 12%–15% of fossil fuel emissions during 1995–2050.[67] Tropical forests can sequester up to 200 metric tons of carbon dioxide per hectare. Organic farming with cover crops can put more carbon back into the soil. Carbon dioxide can be made into useful products by chemical reactions.[68] One reaction is the treatment of carbon dioxide with hydrogen to produce methanol or dimethyl ether for use as fuels.[69] The hydrogen must be obtained from renewable sources (e.g., electrolysis of water using electricity from hydropower, wind energy, or solar energy). If it is made in the usual way by the reaction of methane with water, there is nothing to be gained. Because much more petroleum and natural gas are used for fuels than for the production of chemicals, the significant reduction of carbon dioxide levels by this method would require a major expansion of the chemical industry. Because the chemical industry consumes 22% of the energy used by American industry, it needs to reduce emissions of carbon dioxide.[70] The production of cement generates 10% of the carbon dioxide produced by humans.[71] The worldwide use of energy for refineries is 5 million barrels of oil per day.

Estimates of the cost to prevent further global warming vary widely.[72] A United Nations report on global warming says that the effects of global warming could be mitigated at surprisingly little cost.[73] A study by the Pew Center on Global Climate Change concluded that emissions in developing nations can be reduced without slowing down the countries' economic growth.[74] A group of 2400 scientists feel that the total benefits from reducing emissions will outweigh the cost.[75] They feel that the change can be made without harming employment or living standards. The cost of reducing emissions using some of the advanced technologies now available or on the horizon will be virtually zero, according to two workers in the field.[76] The U.S. Department of Energy (DOE) calculates the net economic cost to stabilize emissions as near zero.[77] The World Resources Institute has estimated that stabilizing carbon dioxide emissions at 1990 levels by 2020 will have a minimal effect on the U.S. economy.[78] The costs of the new technologies will be offset by savings in the fuel that will not have to be purchased. Sir Nicholas Stern calculates the cost to stabilize the concentration of carbon dioxide at a reasonable level by 2050 to be 1% of the gross domestic product.[79]

4.2   HEATING, COOLING, AND LIGHTING BUILDINGS

Buildings produce 39% of the total carbon dioxide emissions in the United States. Sixty percent of the energy used by a building is for heating and cooling, with 40% for lighting and appliances. Reducing the energy used in buildings offers the largest cost-effective way of reducing carbon dioxide emissions.[80] Putting up a new building is best, but retrofits are also good. Before a building can be sold or rented in Europe, its energy efficiency must be calculated and provided to the potential buyers or renters.[81]

4.2.1   USE OF TREES AND LIGHT SURFACES

The large savings in energy that can result from living closer to work in more compact cities[82] was discussed earlier. Cities are islands of heat not only due to the artificial lighting used in buildings, but also due to the many black surfaces that absorb heat from the sun.[83] One-sixth of the electricity used in the United States is used to cool buildings. A combination of lighter roofs, walls, and road surfaces, together with the planting of shade trees in appropriate places, can reduce the need for air-conditioning by 18%–30%.[84] (The range of houses may be as much as 10%–70%, depending on factors such as the amount of insulation in the attic and others). It could make Los Angeles 5°F cooler in the summer. A black roof reflects 5% of solar energy, whereas a white roof reflects 85%. As a white roof becomes dirty, its effectiveness can be reduced by 20%.[85] Washing with soap and water returns its maximum effectiveness. A self-cleansing roof surface might be made with a surface layer of titanium dioxide catalyst for light-catalyzed destruction of the dirt.[86] Another alternative is a surface of a perfluorinated polymer to which the dirt would not adhere well, so that the rain could wash it off. Concrete roads absorb less heat from the sun than ones of black asphalt. Parking lots can be shaded by trees in traffic islands. For individual homes, deciduous trees can be planted on the western and southern sides to cast maximum shadows on the house. A tree on the east side can be used to shade the air conditioner. Shrubs around the foundation will reduce the temperature of the walls and the adjacent soil. Coniferous trees can be planted on the northern side to break winter winds.

4.2.2   SOLAR HEATING AND COOLING

Passive solar heating and cooling[87] can save 50% of the energy needed for a home at the latitude of Philadelphia, Pennsylvania. (A building in temperate, semiarid Argentina that used this technique saved 80% of its heating costs.[88]) This involves putting the long face of the building toward the south. Living spaces can be put on the south and storage spaces and garages on the north. Large windows are put on the south side and small windows on the north side. Other windows are located to allow cross-ventilation in the summer. Overhanging eaves are set to keep out the summer sun, but to allow the lower winter sun to enter the building. The windows should be double or triple glazed with thermal breaks (i.e., insulation) on any metal frames. The building must be well insulated and weather-stripped and follow Leadership in Energy and Environmental Design (LEED, U.S. Green Building Council) specifications. A large mass of concrete, brick, or stone is put in a floor, interior wall, or chimney to store heat overnight. A vent on the roof can be arranged so that heat rising in a stack effect can be used to bring cooler air into the building in the summer. Small turbines used in this way are driven by the rising air or by the wind. Ventilation at night will cool down the building, which, if closed in the morning, will take quite a while to warm up. Heat exchangers can be used with exhaust air to recover some of its heat during the winter.

   Buildings can be heated by putting a transparent plate in front of a wall, preferably blackened, with an air gap in between.[89] The air that heats up in the gap can be moved into the building with a small fan. Such a method can be used to retrofit existing buildings. Solar heat pipe panels have also been used to transfer heat from outside to inside of the building.[90] The water left in the evacuated tube evaporates on the hot outside end and condenses on the cooler end on the inside of the building. Solar roofs have been used to heat air in Northern Ireland.[91] Solar energy has also been used for district heating with the aid of parabolic concentrators that focus the sun's rays on a pipe containing water.[92]

A new solar cooker uses an aluminized plastic sheet parabolic reflector to direct the sun's heat to a storage block beneath the pot inside the house.[93] In another, vacuum tube heat pipes are hooked to the oven.[94] Solar cookers offer a way of compensating for fuelwood shortages in developing countries and of relieving pressure on forests. Wood and crops can be dried faster in a solar tunnel than in the open air.[95] Solar energy has been used to heat swimming pools.[96] Solar-powered water sterilizers used in Hawaii and other places heat the water to 65°C.[97] It is possible to desalinate water with solar energy[98] at half the price of evaporative desalting with fossil fuels or by reverse osmosis.[99]

4.2.3 HEAT STORAGE

The intermittent nature of sunlight requires that there be some means of storing heat at night and on cloudy days. A mass of concrete or masonry is used for that purpose in passive solar homes. The heat can also be stored in beds of rock,[100] in the soil,[101] in tanks of water,[102] or in aquifers.[103] Some of these are for seasonal storage. In one case, 30-m vertical heat exchangers placed in the ground achieved storage efficiencies of 70%. Water pits of concrete lined with stainless steel or high-density polyethylene and insulated with mineral wool have been used in Germany for the seasonal storage of solar heat. One being built in Hamburg, with 3200m² collectors and 4500 m³ of water, is expected to supply 60% of the heat needed in the winter by 120 houses. Another is being built to service 570 apartments. The average temperature for one in Italy on July 31, 1996 was 69.3°C and was expected to reach 80°C by the start of the heating season.[104]

For indoor storage, it is desirable to have storage materials that take up less space, but may cost more. This can be accomplished by materials that undergo phase changes, dehydration, and so on.[105] The compound must change state in a useful temperature range, have a relatively high latent heat per unit volume, and be cheap. Inorganic ones are nonflammable, but may be corrosive. Organic compounds probably would not be corrosive, but could burn. For example, calcium chloride hexahydrate melts at 25°C; thus, it melts by day and freezes by night. But it, along with other hydrates such as magnesium chloride hexahydrate and magnesium nitrate hexahydrate, tends to supercool.[106] The use of 0.5% sodium metaborate dihydrate with a eutectic mixture of magnesium chloride hexahydrate and magnesium nitrate hexahydrate reduced the supercooling from 16°C to 2°C. Another problem is that the anhydrous salt may settle out and not rehydrate properly. This can be minimized using thickeners such as 3% acrylic acid copolymer with sodium sulfate decahydrate containing 3% borax to prevent supercooling.[107] Carboxymethylcellulose (2%–4%) was also used with sodium acetate trihydrate containing 2% potassium sulfate.

4.2.4 LIGHTING

Compact fluorescent lamps use about one-fourth as much energy as the incandescent bulbs that they replace.[108] Such lamp replacement has been a common part of many energy conservation campaigns. For outdoor lighting, low-pressure sodium lamps are eight times as efficient as incandescent lamps, three times as efficient as mercury lamps, and 40% better than high-pressure sodium lamps.[109] The yellow color of the outdoor light should not interfere with most outdoor activities. Astronomers concerned about light pollution recommend "shoebox"-type lighting fixtures that direct the light downward to where it is needed and save energy at the same time.[110] The Czech Republic requires fully shielded light fixtures.[111] A sulfur bulb (the size of a golf ball), which gives off four times as much light as a mercury lamp at one-third the cost, is being tested by the U.S. DOE.[112] The light produced by irradiating sulfur in argon with microwaves is similar to sunlight, but it has very little energy emitted in the infrared and ultraviolet ranges. It puts out as much light as two hundred and fifty 100-W incandescent bulbs. When used with light pipes, it could illuminate large areas of stores, offices, and outdoor areas. The cost of illuminating shopping centers and factories in the United States is US$9 billion each year.

New luminescent materials are being studied for lighting and for displays.[113] A xenon plasma can be up to 65% efficient, compared with low-pressure mercury lamps, which are about 30% efficient. The ultraviolet light from the plasma can be converted to red light with a phosphor in a process during which one photon of vacuum ultraviolet radiation is converted to two visible photons.[114] If suitable phosphors can be found to produce visible light, it could result in the elimination of all ordinary fluorescent lamps that contain mercury. It would increase the efficiency of current xenon lamps used in simulating sunlight in accelerated testing of materials and as sources of light for fiber optic illumination.[115] Workers at Los Alamos Laboratory have made fluorescent lamps in which the mercury has been replaced with a carbon field emitter of electrons.[116] Blue light-emitting diodes (LEDs) based on gallium nitride can be used to produce white light.[117] The combinations of phosphors needed for this are being optimized. They are more energy-efficient than halogen lamps. LEDs are used in traffic lights and tail lights of automobiles. Those that produce white light are available for use in lamps in homes and stores but are still expensive.[118] They offer longer lifetimes and higher efficiencies than conventional light sources. As they become cheaper, it may be possible to light the room with LEDs in the wall. Organic LEDs are also being studied.[119] They have high efficiencies and long lifetimes (in colors other than blue), but the cost of fabrication is high. They are available commercially. These might remove the need for gallium, which is not a very abundant element. The need for the mercury used in fluorescent lamps may be eliminated.

4.3 RENEWABLE ENERGY FOR ELECTRICITY AND TRANSPORT[120]

4.3.1 ALTERNATIVE FUELS[121]

Various alternative fuels for motor vehicles have been suggested as a way of reducing air pollution. Those that are still derived from coal, natural gas, or petroleum will do nothing to mitigate the release of greenhouse gases that lead to global warming.

A bus powered by natural gas is less polluting than that which uses diesel fuel. An electric car[122] transfers the pollution to the power plant where the electricity is generated and where it may be easier to control. If the electricity is made in one of the power plants in the Midwestern United States that was grandfathered by the Clean Air Act and not required to clean up the stack gases, then pollution might be higher with the electric car than with current ones burning gasoline. If the electric cars use lead–acid batteries, lead loss in mining and processing, as well as recycling, must be kept low.[123] If the electricity is made by hydropower, wind power, photovoltaic cells, or other renewable means, then the electric car would reduce both air pollution and global warming. The use of photovoltaic cells on the roof and hood of the car has been suggested.[124]

Dimethyl ether produces less soot and nitrogen oxides when used as a fuel in diesel engines.[125] It may be a way for trucks with diesel engines to meet the particulate standards in the United States. It is made from synthesis gas (Schematic 4.1). The synthesis gas is usually made by treating methane or petroleum with steam at high temperatures. It can also be made from biomass; thus it could help reduce global warming. The methanol made from biomass in this way is a possible fuel for cars. The problem is that at the current level of use of cars and trucks, it would take huge amounts of biomass.

Landfills in the United States produce 8.4 million metric tons of methane each year. The gas from 750 landfills could be recovered economically, but only in 120 was it actually collected in 1995.[126] By 1998, the number had grown to 259.[127] (The landfill in Wilmington, Delaware, will supply gas to the electric power plant a mile away for 20 years).[128] Any carbon dioxide and hydrogen sulfide impurities in landfill gas are removed before the gas is burned for fuel.[129]

Biomass can also be used for energy.[130] In 1999, biomass supplied 3% of the energy used in the United States.[131] Over 2 billion persons in the world rely on biomass fuels.[132] Energy plantations, preferably on land unsuited for other crops, could provide 600 billion kWh of electricity at US$0.04–0.05 kWh^{-1}, according to one estimate. (The prices of electricity quoted here and elsewhere in this chapter come from references of different dates and may not be easy to compare.) Most experiments use fast-growing poplars and willows or switchgrass (*Panicum virgatum*). Switchgrass needs little fertilizer or herbicides and can be harvested twice a year.[133] The European system of coppicing may be suitable for harvesting the biomass from the trees. The tops are cut off every few years, leaving the stumps to resprout. Cutting similar to a hedge might also be suitable. Industrial coppicing for wood for British power plants involves machine harvesting every 3 years.[134] To prevent loss of soil fertility, the ashes from the power plant would have to be spread on the ground of the plantation at intervals. Burlington, Vermont, has a 200 t/day biogasification plant that can use wood, crop residues, yard waste, and energy crops to produce electricity

$$CO + 2H_2 \longrightarrow CH_3OH \xrightarrow[\substack{HMnW_{12}PO_{40} \\ under\ CO}]{225°C} \underset{96\%}{CH_3OCH_3} + \underset{4\%}{CH_3COOCH_3}$$

SCHEMATIC 4.1 Dimethyl ether formation from synthesis gas at US$0.055 kWh[-1].[135] The gas produced by pyrolysis at 1500°F is a mixture of methane, carbon monoxide, and hydrogen. The residual char is used to provide the heat for the pyrolysis. The gases and liquids obtained from biomass by pyrolysis or other means can be upgraded by a variety of catalytic methods, including cracking with zeolites.[136]

4.3.2   A HYDROGEN ECONOMY

Hydrogen has been proposed as a clean fuel that produces only water on combustion.[137] Also, some members of the U.S. Congress have pushed for more research in this area.[138] Almost all hydrogen produced today (96%) is made by steam reforming of coal, natural gas, or petroleum.[139] For a sustainable future, the hydrogen must come from renewable sources.[140] The steam reforming reaction can also be applied to the methane in biogas or in the gas obtained by the pyrolysis of biomass. The water–gas shift reaction can then be run on the carbon monoxide in the pyrolysis gas to produce more hydrogen (Schematic 4.2).

Only 4% of the hydrogen used today is produced by the electrolysis of water.[141] Hydrogen produced in this way costs four times more than that made from petroleum and natural gas. Cheap electricity from renewable sources is needed to encourage the production by electrolysis. The electricity generated by photovoltaic cells can be used to electrolyze water to produce hydrogen. It might cost less and use less equipment if both steps could be run in one apparatus. Photoelectrochemical generation of hydrogen from water[142] has been performed with up to 18% efficiency.[143] It has been carried out with a stacked gallium arsenide–gallium indium phosphide cell with 12.4% efficiency using sunlight.[144] Systems under study include niobates ($A_4Nb_4O_{17}$, where A is calcium, strontium, lanthanum, potassium, or rubidium) used with cadmium sulfide or nickel,[145] cerium(IV) oxide[146] or copper(I) oxide,[147] and titanium dioxide.[148] Dye-sensitized solar cells, using a ruthenium bipyridyl complex with tin and zinc oxides, gave about 8% efficiency in direct sunlight.[149] A seaside power plant in Japan has used solar energy to produce hydrogen from water, then used the hydrogen to reduce carbon dioxide to methane over a nickel–zirconium catalyst at 300°C.[150] The methane was burned to generate more electricity. It would be more direct to generate the electricity with the photocell.

Storage of the hydrogen fuel poses some problems.[151] The gas can be compressed for storage at fixed sites. For use in vehicles, this would add considerable weight in the form of steel tanks. Various methods that would allow more hydrogen to be stored in lighter-weight tanks are being studied. Graphite fiber-reinforced epoxy resin tanks may be suitable. The U.S. DOE target is a

$$CO + H_2O \quad \xrightarrow[\text{350–400°C}]{\text{Fe catalyst}} \quad CO_2 + H_2$$

SCHEMATIC 4.2   Water-gas shift reaction

material that can hold 6 wt% hydrogen at ambient temperature and deliver it in a fully reversible manner over thousands of cycles with short recharging times, preferably without the use of noble metal catalysts.[152] The main classes being studied include metal hydrides, such as ammonia borane (NH_3BH_3),[153] lithium nitrides,[154] and metal organic frameworks, such as zinc dicarboxylates,[155] carbon nanotubes,[156] and aromatic hydrocarbons.[157]

4.3.3 FUEL CELLS[158]

Hydrogen can be burned in a fuel cell to power an electric vehicle. The only product would be water. This is the reverse of the electrolysis of water to produce hydrogen and oxygen. The system is not subject to the Carnot cycle, which limits efficiency in the usual electric power plants,[159] and 60% electrical efficiency can be obtained. Demonstration cars and buses have been built.[160] The fuel cells fit in the same space as the present diesel engine in the bus. An aircraft powered by hydrogen is being planned.[161] Hydrogen is lighter than the present jet fuel, but requires four times as much storage space. The lighter fuel may permit more payload if the tanks that store the hydrogen do not weigh too much. The prototype minivan used tanks made of carbon fiber–reinforced plastic mounted on the roof. It was able to travel 150 miles between refueling. The use of methanol, ethanol, natural gas, or gasoline will require a reformer on board to convert the fuel to hydrogen. Such systems are being studied.[162] Such cars may achieve twice as many miles per gallon as cars with internal combustion engines. A problem that is being overcome is that the traces of carbon monoxide in the hydrogen from reforming poison the catalyst. Fuel cells can now be run directly on butane, ethane, or propane.[163]

The principal types of fuel cells being developed for various uses are (a) molten carbonate, (b) solid oxide, (c) polymer electrolyte, and (d) phosphoric acid. A typical molten carbonate is a mixture of lithium and potassium carbonates at 650°C running at 60% efficiency. If by-product heat can be utilized, then the total energy efficiency will be higher. Fuel cells can run on landfill gas even if carbon dioxide is present. They can have single-cell lifetimes of 30,000 hours.[164] Many cells are stacked in series to obtain the desired voltage. Solid oxide systems, such as zirconium and cerium oxide mixtures,[165] can be run at 600°C or higher. One 25-kW prototype ran 13,194 hours before a scheduled shutdown. No reformer is needed for hydrocarbon fuels. The electrodes do not contain noble metals. A perfluorinated sulfonic acid, such as Nafion, is used as a solid electrolyte in some fuel cells that are the most likely to be used in vehicles.[166] No liquid electrolyte is present. A platinum catalyst for the reaction is used. Concentrated phosphoric acid at 200°C is used in some fuel cells. One in Tokyo ran continuously for 9477 hours before being shut down for scheduled service. Before fuel cells come into widespread use, their cost must be reduced. There is concern that, just as in the case of biofuels, the use of fuel cells has been oversold. Critics feel that greenhouse gas emissions can be controlled more quickly and easily by improving the energy efficiency of vehicles and buildings.[167] The challenge is to get over the idea that each driver has to have his own car and drive alone in it.

4.3.4 Solar Thermal Systems[168]

Heat from the sun can be concentrated by parabolic troughs, parabolic dishes, or by mirrors to produce steam for generating electrical power. These can be used to heat a fluid at the focal point of the parabola from which the fluid is then pumped to the steam generator. The heat can also be aimed at a central receiver. The central receiver allows higher temperatures to be reached. These systems can be made to track the sun for greater efficiency. A 750-ha plant in the California desert using parabolic troughs has operated for 20 years with more than 94% power availability. It produces electricity for as little as US$0.10 kWh^{-1}. Further improvements, such as vacuum insulation for the heat-carrying pipes, better absorbers, antireflective coatings, and such, are expected to lower the cost of the electricity to a level at which it can be competitive with that from conventional plants that burn fossil fuel.

Solar absorbers of Mo/Al_2O_3 can be deposited on copper-coated glass using two separate electron beams.[169] These offer 0.955 absorbance at 350°C, when topped with an antireflective coating of alumina. Another offers 90% solar absorbance with low thermal emittance through the use of a 50-nm Al_2O_3/10–12-nm AlCuFe/70-nm Al_2O_3-layered coating.[170] The AlCuFe coating is applied by ion beam sputtering. Cathodic arc chemical vapor deposition of absorbent surfaces is reported to be faster than other methods.[171] The stainless steel–aluminum nitride coatings put on by magnetron sputtering (mentioned earlier) are stable to 330°C–400°C.[172] Black cobalt oxide absorbers applied by electrodeposition are stable to 300°C, but not to 400°C.[173] Further research is directed at absorbers that will be stable at 500°C.

4.3.5 Photovoltaic Cells[174]

Photovoltaic cells convert light to electricity at the place where it is to be used. There is no need for a central facility with a concentration of equipment or for long transmission lines (as well as the losses of electricity in them). The peak in the electricity generation may well occur during a hot summer afternoon when the electrical system of the community is strained and expensive supplemental generators must be brought into service. Thus, its cost should be compared with the cost of this supplemental electricity in this situation. (In 2009, electricity from the best photovoltaic systems cost about twice as much as peak power from a conventional plant.) Photovoltaic systems are cost-competitive now in places remote from power lines. Further research is needed to make them cost-competitive in areas where an electric network is already in place. The current problem is that the most efficient photovoltaic cells are expensive to produce and the cheaper ones are not efficient enough. Photovoltaic systems use no toxic fuel, use an inexhaustible fuel, produce no emissions, require little if any maintenance, and have no moving parts. In 1998, the United States produced 44% of the world's solar cells and Japan 24%.[175] In 2004, the United States produced 12%, Japan 50%, Europe 26%, and the rest of the world 12%.[176] The market for photovoltaic cells was growing at the rate of 40%/yr in 2005.[177]

High-efficiency solar cells can be made from silicon wafers cut from single crystals grown from a melt.[178] One of the most efficient systems is made by a series of steps.[179] Its efficiency is 23.5% under one sun. (one sun means that no mirrors are used to concentrate the light. Because mirrors are cheaper than solar cells, they are sometimes used.)

The number of steps needs to be reduced to reduce the cost.[180] One proposal is to coat both sides at once.[181]

Uncoated, untextured silicon may reflect as much as 36% of the incident light. This can be reduced to about 12% by a single coat of titanium put on by the sol–gel method, and to 3% by a double titanium dioxide coating.[182] Other materials for antireflective layers include silicon nitride,[183] tantalum oxide,[184] radiofrequency-sputtered indium tin oxide,[185] and zinc oxide.[186] Aluminum nitride topped with three layers of angled titanium dioxide nanorods produced a coating that reflected almost no light.[187] Nanoporous poly(methyl methacrylate) allows 99.3% of the incident light to pass through.[188] A sulfur hexafluoride–oxygen plasma has also been used to texturize silicon surfaces.[189] Abrasive cutting wheels have also been used.[190] The discoloration of the ethylene–vinyl acetate encapsulant can be reduced by stabilizers or by the use of a glass that filters out the ultraviolet light.[191]

Cutting the silicon wafer into thin slivers increases the surface area 13-fold to produce a cell that is about 20% efficient.[192] The energy to produce the cell can be recouped in 1.5 years compared with 3 years for a conventional cell. Polycrystalline silicon is cheaper than single-crystal silicon. It can be made by slow casting of an ingot of silicon. An 18.6% efficiency has been obtained in a cell made with it.[193] It is made by reduction of trichlorosilane with hydrogen.[194] Amorphous silicon has a higher absorbance than crystalline silicon and can be used in thinner layers.[195] It can be crystallized with a laser, but this might require application of several layers to obtain the thickness needed for a solar cell.[196] Amorphous hydrogenated silicon can be made in a plasma from silane and methane.[197] Such thin-film cells[198] can reach efficiencies of 9%–10%. Rear reflectors increase efficiency by keeping the light in.[199] Unfortunately, these thin-film cells have a lower efficiency after a few years owing to the diffusion of aluminum from the contacts into the silicon. Some of the loss can be restored by annealing at 150°C. Some of the loss can also be avoided by using thermal insulation to keep the cell hotter.[200] If a triple junction cell (i.e., one containing the equivalent of three stacked cells) is used, the efficiency can be as high as 12.8%.[201] Flexible plastic cells of amorphous silicon have been made by automated roll-to-roll chemical vapor deposition processes.[202] Amorphous silicon on crystalline silicon has been used in an effort to decrease the thickness of crystalline silicon needed.[203] The cell efficiency was 14.7%–16%.

The most efficient photovoltaic cell (42.8%) uses multiple junctions of indium gallium phosphide, gallium arsenide, silicon, and indium gallium arsenide.[204] The next best (40.8%) uses a triple junction of gallium indium phosphide, gallium arsenide, and germanium at 326 suns.[205]

A group in Switzerland has developed a photovoltaic cell that can function as a window for a building.[206] In one example, a ruthenium pyridine complex photosensitizer is attached to the titanium dioxide semiconductor by a phosphate. An iodine-based electrolyte (KI_3 dissolved in 50:50 ethylene carbonate/propylene carbonate) is between the panes. All the films are so thin that they are transparent. The efficiency is 10%–11%. Other dyes have been tested to avoid the use of expensive ruthenium.[207]

Silicon has the advantage of being abundant and nontoxic. It is used in 90% of solar cells. Most other cells contain toxic metals or elements that are not abundant and probably not cheap. If materials other than silicon come into common use, some sort of collection and recycling system will be needed to keep discarded cells from ending up in incinerators and landfills where they may be sources of pollution.[208] (Silicon Valley, California, has 29 Superfund sites.[209])

Solar cells are being used for many purposes in many places around the world: for heating and lighting homes, lighting bus stops, pumping water, and many others.[210] They can be used in place of the usual cladding or roofing of a new building, which avoids some of the costs for the usual materials used for these purposes.[211] They may work best if cleaned and serviced at intervals by an independent serviceman or one connected with the local utility.[212] Individual home owners may forget to clean them. Self-cleaning modules based on the photocatalytic action of thin layers of titanium dioxide may be possible.[213] A home in Osaka, Japan, fed 77% of its photovoltaic power into the grid between 10:00 a.m. and 2:00 p.m. on a clear day in May.[214] This suggests that 3 kW of photovoltaic power in a house at this location would satisfy most of the needs of the house and alleviate peak power demand. A house in Florida with photovoltaic cells and white roof tiles required 83% less energy to air-condition than a conventional house.[215] With more solar cells, it might not require any energy from the electrical grid. It is also possible to utilize both the heat and light of the sun with the solar array.[216]

4.3.6 OTHER SOURCES OF RENEWABLE ENERGY

Other sources include hydropower, tidal energy, wave energy, geothermal energy, and wind.[217] Further development of hydropower[218] is possible, for many existing dams do not involve the generation of electricity. This may be partly due to the small heads of water involved. The erection of new dams for this purpose involves tradeoffs, because natural biotic communities as well as agricultural fields, houses, and towns are often lost in the process. The wind was used more as a source of power in the past than it is today. Before the days of rural electrification in the United States, it was common practice to use windmills to pump water for the farm. The Netherlands is famous for its windmills of former years that were used to grind grain, and so on. Windpower is cost-competitive today in some areas.[219] Part of the cost is offset because the land under the turbines can be used for other purposes, such as farming and grazing. Some new projects are producing electricity at a wholesale price of US$0.045 kmh^{-1}.[220] Wind farms are located in places where the wind blows fairly steadily, such as mountain passes and at the seashore. The Great Plains of the western United States has been suggested as a place for wind farms. A 100 MW wind farm is planned for Minnesota and a 50 MW one for Iowa. The power of the wind goes up as the cube of the wind speed. In 1994, California had 15,000 wind turbines, generating enough electricity to supply a city the size of San Francisco. They operated 95% of the time. The wind turbines in Denmark supplied 20% of the country's electricity in 2006.[221]

Electricity is generated from steam in the ground in places such as New Zealand, California, Nevada, Utah, and Hawaii.[222] Geothermal energy is also used to heat 80% of the homes in Iceland. The steam and hot water from underground reservoirs contain gases, such as hydrogen sulfide and ammonia, as well as dissolved minerals that may contain arsenic, mercury, vanadium, nickel, or others. The geothermal plants can be polluted if these are discharged at the surface. A better system is to reinject the used water, not only to avoid the need to treat the water to remove pollutants, but also to form more steam and to prevent land subsidence. A plant is being built in the Imperial Valley of California to recover 30,000 metric tons of zinc annually before the reinjection of the used water.[223] The plants at the geysers near Santa Rosa, California, have been seeing steam production drop 10%/yr since the mid-1980s, despite reinjection of 31% of the water produced.[224] A better system is to use a closed loop with a working fluid other than water (e.g., isobutene or pentane). This may also be better in utilizing low-grade heat. Studies have also been made of using the heat of hot, dry rock to produce steam to generate electricity. The temperature at the bottom of a 12,000-ft borehole in New Mexico was 177°C (350°F). For production of steam, it will probably be necessary to shatter the rock at the bottom of the hole with explosives. Oil well fracturing techniques can also used. A 1 MW system of this type is being built in Australia.[225]

Tidal energy can be used to generate power in places where the tidal range is fairly large.[226] Power is generated as the water goes in over a dam and again as the water comes out over the dam. A plant in France uses this method. Proposals have been made to use the exceptionally high tides in the Bay of Fundy in Canada for this purpose. The waves in the ocean are a much more general source of power, which is largely untapped. The advantages of waves over offshore wind power include that waves occur more of the time than wind, there is no damage to birds and bats, and there is minimal visual impact for people to complain about. A few prototypes have been tested for the generation of power. The best may be those with the fewest moving parts. One system uses sand-filled ballast tanks to keep the apparatus in place.[227] As the wave passes, air is expelled upward through turbines that generate electricity. As the water level drops, another air turbine will generate electricity. A wind turbine can be mounted on top, if desired. The electricity from such a unit in Japan costs about twice as much as that from an oil-fired power plant.[228] Other schemes use pistons actuated by the water to drive a generator. The up-and-down motion of a floating device anchored to the bottom (a "bobbing duck") could also be used to generate power. An underwater system utilizes the pressure differences caused by the passing of the wave. The air in two mushroom-shaped floaters that are each partially filled with water is transferred between the two as the wave passes.[229]

The challenge is to reduce the costs of the various forms of renewable energy[230] to where they can compete with fossil fuels today. If the various environmental and health costs were included in the price, the cost of electricity from an unscrubbed coal-burning power plant would about double. This new form of accounting would make it easier to replace fossil fuels with renewable forms of energy.

4.3.7 Energy Storage

Because some of the sources of renewable energy are intermittent, it will be necessary to combine them with some that are not, or find ways to store enough energy until the source is again available. It might be possible to supplement energy from the wind or the sun with energy from biogas, biomass, or hydropower. This might require building an additional plant and the expense that goes with it. At present, solar and wind power are supplemented with that from natural gas and other fossil fuels or with hydropower.

Energy can be stored using pumped storage reservoirs, compressed air in tanks, flywheels, or hot water.[231] Flywheels with electromagnetic bearings can be up to 90% efficient in storing and releasing energy.[232] The possibility of storing energy in highly extendible materials, such as rubbers, is being studied.[233] An additional possibility is to use light to form a higher-energy compound that can later be reconverted to the original compound with the release of the energy. The reaction must be 100% reversible over many cycles, should work with sunlight, and should have a high quantum yield. The most studied reactions of this type involve the conversion of norbornadienes to quadricyclanes, which may be small molecules or attached to polymers.[234] A typical example is shown in Schematic 4.3.

Storage of energy in batteries[235] is common. The search is on for batteries of common nontoxic materials that combine light weight and compactness with high-energy density. The most common goal is a better battery for widespread use in motor vehicles.[236] For this use, it must be fully reversible over many cycles and be relatively easy to recharge quickly. The major candidates for electric vehicle batteries are shown in Table 4.1.

The first three are commercially available today. The nickel–cadmium battery has the disadvantage of using toxic cadmium. The lead–acid batteries used in cars today weigh too much and wear out too soon for widespread vehicles. The nickel–metal hydride battery appears to be available today for general use.[237] It is essentially a nickel–hydrogen cell with a hydrogen-storing alloy, such as $LaNi_5$, as the anode with aqueous potassium hydroxide containing some sodium hydroxide and lithium hydroxide as the electrolyte.[238] The active material in the cathode is nickel

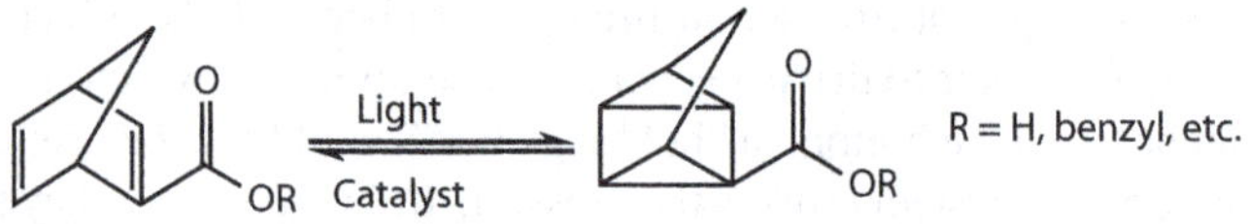

R = H, benzyl, etc.

SCHEMATIC 4.3   Formation of quadricyclanes from norbornadienes

TABLE 4.1
Candidates for Electric Vehicle Batteries

Battery	Range/Charge (miles)	Life (years)	Specific Energy (W-h/kg)
Nickel–cadmium	65–140	8	45–55
Lead–acid	50–70	2–3	35–40
Nickel–metal hydride	100–120	8	High 60s–70
Lithium ion	150–300	5–10	100–150

$$Ni(OH)_2 + M \rightleftharpoons NiOOH + MH$$

SCHEMATIC 4.4 Nickel-metal hydride battery reaction

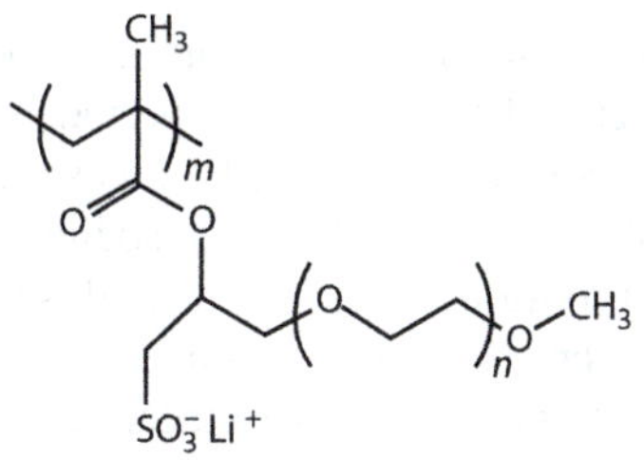

SCHEMATIC 4.5 Lithium chlorate-polymer combination

hydroxide. The overall reaction in the battery is shown in Schematic 4.4 (where M represents the hydrogen-storing alloy).

Lithium batteries are being studied intensively owing to their light weight and high energy density.[239] Modifications to prevent the occasional fire when overcharging laptops are being made. Most lithium batteries use a lithium-intercalated carbon anode and a cathode of a lithium metal oxide made of cobalt, nickel, or manganese. Lithium iron phosphate is thought to be safer. It is being used in portable power tools and in a car battery made by A123 for an electric car maker in Norway.[240] The electrolyte is a lithium salt, such as perchlorate, hexafluorophosphate, hexafluoroarsenate, or triflate, in solvents such as ethylene carbonate and propylene carbonate. The electrolyte can also be a solid containing the salt, such as polyethylene oxide or polypropylene oxide, which yields an all-solid battery.[241] A plasticizer can be added to increase conductivity in the solid electrolyte. The challenge is to increase the conductivity to the desired range of 10^2–10^1 S/cm. An electrolyte film of a poly(epichlorohydrin-co-oxirane) and poly(acrylonitrile-co-1,3-butadiene) blend, swollen with a solution of 40% lithium perchlorate in propylene carbonate, has a conductivity of more than 10^3 S/cm.[242] A combination of polymer (Schematic 4.5) with lithium chlorate has a conductivity of 10^3 S/cm.[243]

4.4　USE OF LESS COMMON FORMS OF ENERGY FOR CHEMICAL REACTIONS

4.4.1　ELECTRICITY

Electricity is used on a large scale in the production of inorganic compounds such as chlorine and sodium hydroxide. It is used to refine metals such as aluminum and copper and can also be applied to titanium.[244] However, the only large-volume application in organic chemistry appears to be the hydrodimerization of acrylonitrile (Schematic 4.6) to form adiponitrile (2,000,000 tons/yr), which is then reduced to the diamine or hydrolyzed to adipic acid for the preparation of nylon-6,6.[245]

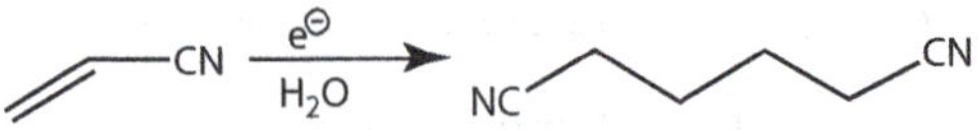

SCHEMATIC 4.6 Hydrodimerization of acrylonitrile

Electrochemistry[246] can be used to regenerate an expensive or toxic reagent *in situ*. An example is the electrical regeneration of periodate that is used for the oxidation of vicinal diols such as sugars. When used with osmium oxidations, it can keep the level of the toxic metal reagent quite low. The sodium hydroxide and iodine produced by electrolysis react to form hypoiodite that adds to the double bond to form the hydroxyiodide, which then eliminates sodium iodide by the action of the sodium hydroxide to reform the sodium iodide.[247] Electricity can be used in oxidations and reductions instead of reagents. When iodine is reduced to hydriodic acid in this way, no waste products are formed.[248] Phosphorus-containing wastes are produced in the conventional process using phosphorus. The new process is also cheaper. Ketones can be converted to alcohols in up to 98% yield by electrocatalytic hydrogenation using rhodium-modified electrodes.[249] There is no catalyst to be separated at the end of the reaction. The electrochemical hydrogenation of water-immiscible olefins and acetylenes is enhanced by concurrent ultrasonication.[250] Hydrogenation of edible oils using a cell with a Nafion membrane with a ruthenium anode on one side and a platinum or palladium black cathode on the other produced less of the undesirable *trans*-isomers than conventional hydrogenations.[251]

4.4.2 Light

Light is environmentally benign, leaving no residues to be removed in the workup of a reaction. It can catalyze some reactions that are difficult or impossible to run in other ways (e.g., some cycloadditions).[252] Most photochemistry is carried out using light with shorter wavelengths than those found in sunlight. The challenge is to use sunlight as is, or when concentrated. The degree of concentration can vary with the optical system. The large concentration obtained with solar furnaces has been suggested for isomerization, cycloadditions, catalytic cyclizations, and purification of water.[253]

A few examples will be given to illustrate current trends in research. Dithianes, benzyl ethers, and related compounds have been cleaved by the use of visible light with a dye (Schematic 4.7).[254] (To use more of the light, mixtures of dyes can be used.)

Benzonitrile in methanolic potassium hydroxide can be hydrolyzed to benzamide in 96% yield using an oxophosphoporphyrin catalyst with 420-nm light at 20°C.[255] A combination of visible light and water was used in a cyclization (Schematic 4.8) to produce substituted pyridines with almost no by-products.[256] Pinacols have been made with sunlight (Schematic 4.9).[257]

Epoxides can be polymerized with visible light.[258] Oxidation of hydrocarbons in zeolites with blue light gives improved selectivity.[259] Isobutane can be converted to *tert*-butylhydroperoxide with 98% selectivity. Benzaldehyde is produced from toluene, acrolein from propylene, and acetone from propane.

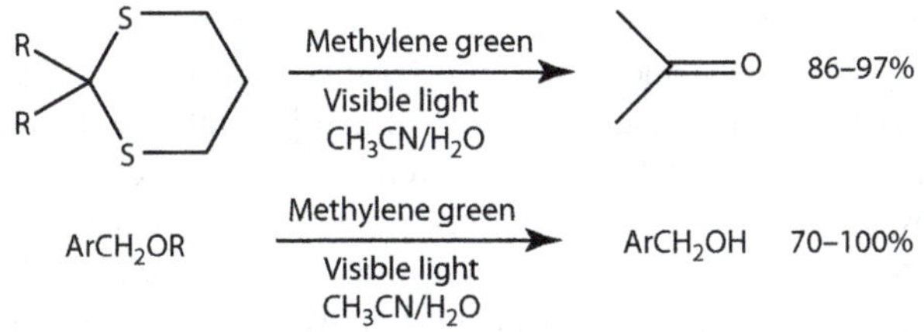

SCHEMATIC 4.7 Light cleavage of dithianes and benzyl ethers

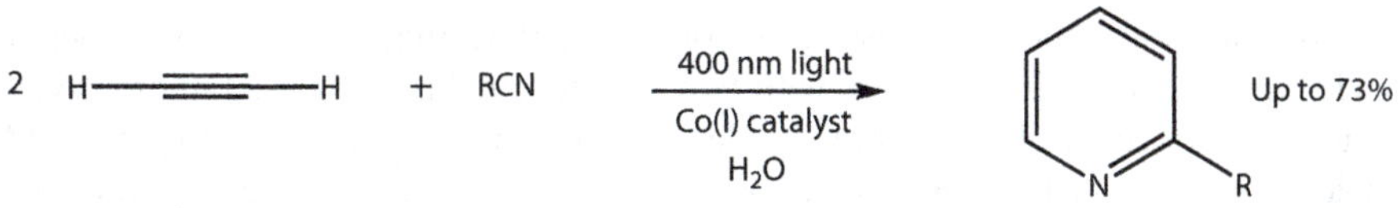

SCHEMATIC 4.8 Pyridine synthesis

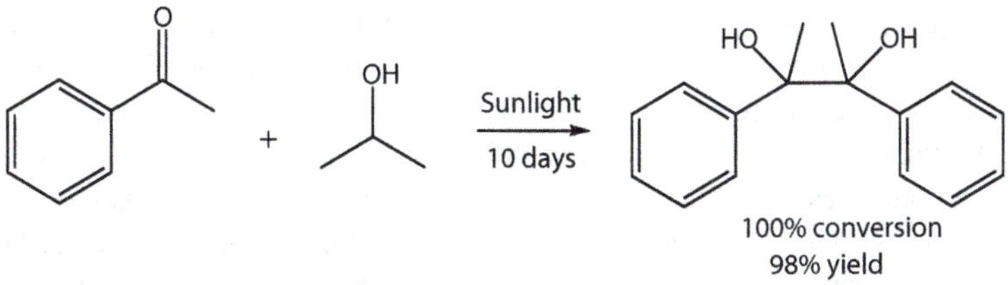

SCHEMATIC 4.9 Pinacol synthesis using sunlight

4.4.3 ULTRASOUND

This technique is especially useful when a solid in a liquid has to react. This may be a matter of the surface being cleaned continuously by the cavitation. (Cavitation produces high temperatures and pressures for very short times.) It can help when there are two immiscible liquids. In this case, mechanical stirring for macromixing may be combined with ultrasound for micromixing. The instantaneous high pressures obtained can be used to accelerate some reactions when the products occupy a smaller volume than the starting materials (e.g., the Diels–Alder reaction).

The cost of the ultrasonic energy needed for these mixing and cleaning processes in reactions is not high because only relatively small amounts of energy are needed. In contrast, the energy needed for doing chemistry with ultrasound is much higher (e.g., generating hydroxyl radicals in water). If the hydroxyl radical is to be used to initiate a polymerization, only a little may be needed, and so it may not cost too much. The cost is still not prohibitive for reactions such as lowering the molecular weight of polymers, if relatively few chain scissions are needed. A couple of examples from the field of polymers will illustrate some of what can be done. Block copolymers of polyethylene and acrylamide have been made by ultrasonic cleavage of the polyethylene in the presence of acrylamide.[260] Maleic anhydride grafted to polypropylene in 93% yield when the two were ultrasonicated at 60°C in the presence of benzoyl peroxide.[261] Nitration of phenol with 9 wt% aqueous nitric acid can be done in 2 hours with a 94% conversion to a 70:27 *p*-nitrophenol/*o*-nitrophenol mixture with ultrasound but the conversion is only 30% to a 49:48 *p*-nitrophenol/*o*-nitrophenol mixture in 48 hours without it.[262]

4.4.4 MICROWAVES

Use of microwave energy,[263] instead of conventional heating, often results in better yields in very short times. Ovens with homogeneous fields of focused microwaves, if temperature control and power modulation are present, can lead to improved yields compared with conventional ovens. Such ovens are supplied by Biotage, CEM, Milestone, and Anton Paar. There must be something present that absorbs the microwaves.

In the synthesis of zeolites, it is necessary to hold the gel at elevated temperatures for extended periods to induce the zeolite to crystallize. Zeolites can be made from fly ash in 30 minutes when microwaves are used, compared with 24–48 hours when the mixture is just heated.[264] In the synthesis of zeolite MCM-41, heating for 80 minutes at 150°C is replaced by the use of microwaves for 1 minute at 160°C.[265] Varma et al. have run many organic reactions with clay catalysts without solvent in open vessels for 1–3 minutes to obtain good yields of products.[266] These include the preparation of imines and enamines from ketones and amines in 95%–98% yields, oxidation of alcohols to ketones with iron(III) nitrate on clay, cleavage of thioketals, and others. The rearrangement of a cyclic oxime to a lactam has been carried out in quantitative yield with microwaves in 5 minutes in the presence of the zeolite AlMCM-41 (Schematic 4.10).[267]

That these yields cannot be obtained in a comparable period of simple heating make these workers feel that there is a special microwave effect. It is possible that the effect is due to instantaneous high temperatures at the surface of the solid. It may also be that microwaves facilitate contact between the solid and liquid reagents. If so, a combination of ultrasound and microwaves might be even better for reactions that take somewhat longer to run with microwaves than those just cited. Others feel that there is no such effect.[268] While the debate about the special microwave effect is still going on, it appears to have been settled for at least this one class of reactions.

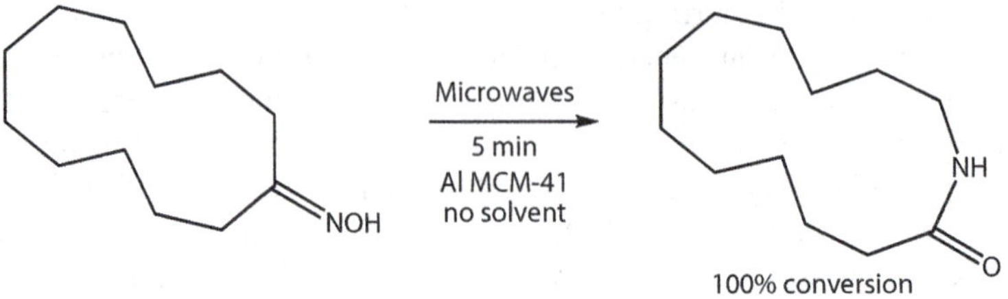

SCHEMATIC 4.10   Microwave-initiated Beckmann rearrangement

The application of microwave energy as an alternative to conventional heating needs to be tried wherever microwave-absorbing (i.e., polar) materials are present. These include not only inorganic compounds, but also water, alcohols, amides, and many other oxygen-containing compounds, but not hydrocarbons, such as hexane and toluene. Xylene can be heated with microwaves by dispersing 1%–2% of cobalt or magnetite, 5–2-nm nanoparticles, in it.[269] A magnet is used to recover these for reuse in the next run. Microwaves enhance the solid-state polymerization of polyethylene terephthalate at 236°C and nylon at 182°C–226°C by enhancing the diffusion of small molecules.[270] There is considerable current enhancement when microwaves are used with platinum microelectrodes.[271] Microwaves have been used with metal powders at high temperature to produce parts that are superior to those made by the usual sintering in powder metallurgy.[272] They have also been used in the production of ceramics.[273] Regeneration of zeolites that had been used to trap volatile organic compounds with microwaves required only one-third as much energy as conventional heating.[274]

Microwaves should be well suited for continuous tubular reactors, where they might reduce the amount of energy needed, reduce costs, and increase the output of a plant. If water were inert, it might be used to absorb the microwave radiation, the reagents being in solution or in an emulsion. Further research will undoubtedly find many more applications for this technique.

PROBLEMS

4.1 A Sticky Issue

Imagine that you are a university architect who is designing new buildings that are expected to be more energy-efficient. This will help in slowing climate change due to increasing levels of carbon dioxide in the air (now at 400 ppm). Passive solar heating and cooling can save half of the energy needed to heat and cool a building at half the cost of solar arrays. The state of California requires it in new and retrofitted buildings. It can also be combined with solar arrays to approach the goal of the "zero energy house." Passive solar heating and cooling adds about 10% to the cost of the building, but the payback period is short—in fact, much shorter than that for solar arrays. You have created some great building designs, but the vice president of finance will not accept them because money is tight and he does not want salaries to be cut. What can be done to change his mind?

4.2 A Nuclear Fantasy

The vice president of public relations for Second Energy (one of the nation's largest utilities) made the following remarks to the Nuclear Regulatory Commission when applying for permits to build five nuclear plants in northern Delaware:

> The new plants will feature the latest technology in pebble bed reactors (containing no sand, gravel or rocks). The neutrons will be piled higher and deeper. The electrons will be shipped promptly, all in a fail-safe mode. The plant will be operating 95% of the time. If two reactors are bought, availability will go up to 100%. The solid by-products will be stored on the ground in concrete casks until the time when a reprocessing plant will be built. The waste heat will be used to power buildings in the cities, e.g., the University of Delaware campus.

The plant will be surrounded by a double fence topped with razor wire and with constant surveillance. There will be double checkpoints where fingerprints and eye color will be monitored. Vacant land inside the fence will be used for a fish hatchery and an employee golf course. Some waste heat might additionally be used for a greenhouse where tomatoes can be raised for employees. The new plants will help to combat global warming. They will not have the stigma associated with emissions of carbon dioxide.

Communities that want to grow and prosper will welcome the plants for their distinctive architecture, as well as the jobs and money that they will bring. The reliable supply of electricity will attract other industries. Unfortunately, there will not be enough plants so that every community can have one. Towns might have to provide some tax incentives to induce a company to build one.

What is wrong with this scenario?

4.3   A Hectic Life

Your job is that of a principal investigator at the Old Glory Chemical Company. Your current work deals with a new light source that contains no mercury, and is tremendously more energy-efficient than a standard fluorescent lamp. Recently, you seem to be spending more time justifying your work than doing it. The problem started when investors wanting a higher dividend ousted the chief executive officer and replaced him with an accountant from outside the company. The job today is to prepare a report on why you spent so much money buying a new rotary evaporator last month. The chances of replacing the beaten-up old tube furnace seem remote.

You need to prepare for an important meeting that will get more money for the project. The vice president for research wants to hear about the work on the new lamp. Your supervisor is so apprehensive that today you have to go through the third dry run for the meeting. So much chemistry has been deleted and so much hype has been added that you can hardly recognize your work. Then, there is the paper for the trade journal that the Advertising Department wants you to publish. It is even worse, now in its fifth revision. Next week, there is the meeting with The Lamp Company, which does the actual testing of the phosphors and filaments that you design. You are supposed to get good test results without revealing too much chemistry. If their questions get too factual, the answers have to be evasive, especially since a representative of the Patent Department will be there. Then, there is the mass spectrometry laboratory that you have to depend on. The overworked chemist there (since the recent downsizing) seems to do the samples that interest him most, instead of doing the urgent ones first.

The biggest problem is with your technician. He received a B.A. in education, but after a year with a local school district he was unfortunately laid off. He is bright and does a good job. However, after a year with you, he feels that he knows a lot more chemistry than he really does and wants to design and run his own energy-related experiments, which are often separate to your current program. You fear for his safety and your safety, as he wants to work with some very dangerous materials. You still have not finished filling out the forms and giving him the additional safety training, required by the accident with the electrical socket last month. How can you motivate him and regain his confidence so that his view of his chemical knowledge is more realistic?

4.4 What We Know How to Do but Are Not Doing

There are many activities that we could do to reduce the demand for energy but do not. These might also save money and lower the cost of health care. Make a list of as many as you can think of. Is the initial investment too high? Is the payback period too long? Does change bother us? Are we too comfortable with the status quo?

4.5 An Episode of Fracking

There is currently a boom in natural gas in the United States. The chemical industry is expanding due to this, which has created new jobs and increased competitiveness with other nations. Natural gas is viewed as a bridge from current greenhouse gas generation to a future with renewable energy. The boom is a result of the development of horizontal drilling and hydraulic fracturing of rock around 1500 m below the Earth's surface (known as "fracking"). However, not everyone is happy about this situation. There are stories about methane coming out of the water tap and burning when lighted, prompting consumers to switch to bottled water. Water from the wells can additionally contaminate rivers. Imagine that you are the scientist at the Environmental Protection Agency who has to set new regulations to assure that fracking is done in an environmentally responsible way. How can you tell which, if any, of the horror stories are real? What regulations are required?

4.6 Can More Material Goods Lead to Happiness?

Americans often try to find happiness by buying more and more items, leading to increased production costs and energy usage. This overconsumption is an unsolved problem. Does everybody really need a bigger house or car? Are people trying to impress their neighbors with their wealth? Golf courses are not as popular as they used to be. One in Wilmington is sprouting million dollar houses that are selling well. Suggest some possible ways to change this mindset to save energy, materials, and money.

REFERENCES

1. (a) J. Randolph and G.M. Masters, *Energy for Sustainability—Technology, Planning, Policy*, Island Press, Washington, DC, 2007; (b) J. Goldemberg, ed., *World Energy Assessment—Energy and The Challenges of Sustainability*, United Nations Development Programme, New York, 2001; (c) E. Kintisch, *Science*, 2007, *318*, 547; (d) J. Holdren, ed., *Science*, 2007, *315*, 737–812 (issue on energy); (e) N. Armaroli and V. Balzani, *Angew. Chem. Int. Ed.*, 2007, *46*, 52–66; (f) L.R. Brown, *Futurist*, 2006, *40*(4), 18; (g) L.R. Brown, *Plan B 4.0, Mobilizing to Save Civilization*, W.W. Norton, New York, 2008; (h) Worldwatch Institute, *American Energy—The Renewable Path to Energy Security*, Washington, DC, 2007; (i) Union of Concerned Scientists, www.ucsusa.org/clean_energy; (j) R. Dell and D. Rand, *Clean Energy*, Royal Society of Chemistry, Cambridge, 2004.
2. A. Vijayan and A. Kumar, *Environ. Prog.*, 2005, *24*(2), 125.
3. A. Bisio and S.R. Boots, *Energy Technology and the Environment*, Wiley, New York, 1995, 1131.
4. Anon., *Chem. Eng. News.*, 1996, May 17, 23.
5. R. Winder, *Chem. Ind. (Lond.)*, July 4, 2005, 14.

6. (a) C.A.C. Sequeira and J.B. Moffat, eds, *Chemistry, Energy and the Environment*, Royal Society of Chemistry, Cambridge, 1998; (b) J. de S. Arons, H. van der Kooi, and K. Sankaranarayanan, *Efficiency and Sustainability in the Energy and Chemical Industries*, Marcel Dekker, New York, 2004; (c) W.J. Storck, *Chem. Eng. News*, Feb. 5, 2001, 19.

7. (a) R.E. Hester and R.M. Harrison, eds, *Environmental Impact of Power Generation*, Royal Society of Chemistry, Cambridge, 1999; (b) R.F. Hirsch and A.H. Serchuk, *Environment*, 1999, *41*(7), 4.

8. Anon., *Wilmington Delaware News J.* 1996, Aug. 13, A5.

9. J.I. Levy, J.K. Hammitt, Y. Yanagisawa, and J.D. Spengler, *Environ. Sci. Technol.*, 1999, *33*, 4364.

10. Anon., *Environ. Sci. Technol.*, 1998, *32*, 399A.

11. Anon., *Chem. Eng. News*, 1999, Jan. 11, 22.

12. R. Tyson, *Environ. Sci. Technol.* 1997, *31*, 508A.

13. J.A. Cusumano, *J. Chem. Ed.*, 1995, *72*, 959.

14. R.L. Berglund, *Kirk–Othmer Encyclopedia of Chemical Technology*, 4th ed., Wiley, New York, 1994, *9*, 1022.

15. (a) L. Ruber, *Chem. Eng. News*, 1997, Oct. 20, 14; 1998, Oct. 5, 11; (b) W.V. Cicha, *Chem. Eng. News*, 1997, Dec. 15, 6; (c) G. Parkinson, *Chem. Eng.*, 2000, *1072*(2), 27.

16. J.J. Corbett and P. Fischbeck, *Science*, 1997, *278*, 823.

17. J. McCormick, *Acid Earth—The Politics of Acid Pollution*, 3rd ed., Earthscan, London, 1997.

18. Anon., *Environ. Sci. Technol.*, 1997, *31*, 459A.

19. G. Davison and C.N. Hewitt, *Air Pollution in the United Kingdom*, Royal Society of Chemistry, Cambridge, Special Publication 210, 1997.

20. Anon., *Environ. Sci. Technol.*, 1998, *32*, 171A.

21. (a) G.E. Likens, C.T. Driscoll, and D.C. Buso, *Science*, 1996, *272*, 244; (b) G.E. Likens, K.C. Weathers, T.J. Butler, and D.C. Buso, *Science*, 1998, *282*, 1991; (c) J.M. Gunn, W. Keller (and following papers), *Restor. Ecol.*, 1998, *6*, 316–390.

22. Environmental Action Fact Sheet, Washington, DC, Feb. 1991.

23. Environmental Defense Fund, Jan. 2002.

24. (a) Anon., *Chem. Eng. News*, 1994, Sept. 19, 20; (b) C. Steiner, *$20 Per Gallon*, Grand Central Publishing, New York, 2009; (c) A. Schafer, H.D. Jacoby, J.B. Heywood, and I.A. Waltz, *Am. Sci.*, 2009, 97, 476; (d) D. Sperling and D. Gordon, *Two Billion Cars: Driving Toward Sustainability*, Oxford University Press, New York, 2008.

25. J.J. Romm, *Hell and High Water: Global Warming—The Solution and the Politics— What We Should Do*, William Morrow, New York, 2007.

26. G. Peaff, *Chem. Eng. News*, 1997, Aug. 4, 20.

27. H.J. Hebert, *Wilmington Delaware News J.*, 1997, Oct. 18, F5.

28. P.H. Abelson, *Science*, 1992, *257*, 1459.

29. P.H. Abelson, *Science*, 1997, *227*, 1587.

30. T. Krawczyk, *Int. News Fats Oils Relat. Mater.*, 1996, *7*, 1320.

31. (a) T.R. Loughlin, ed., *Marine Mammals and the Exxon Valdez*, Academic, San Diego, 1994; (b) D.A. Wolfe, M.J. Hameedi, J.A. Galt, G. Watabayashi, J. Short, C. O'Claire, S. Rice, J. Michel, J.R. Payne, J. Braddock, S. Hanna, and D. Sale, *Environ. Sci. Technol.*, 1994, *28*, 560A; (c) A. Weiner, C. Berg, T. Gerlach, J. Grunblatt, K. Holbrook, and M. Kuwada, *Restor. Ecol.*, 1997, *5*, 44; (d) S.F. Sugai, J.E. Lindstrom, and J.F. Braddock, *Environ. Sci. Technol.*, 1997, *31*, 1564; (e) S.D. Rice, R.B. Spies, D.A. Wolfe, and B.A. Wright, eds, *Proceedings of the Exxon Valdez Oil Spill*, American Fisheries Society, Bethesda, MD, 1996; (f) *Chem. Eng. News*, 1994, Aug. 1, 17; (g) R.M. Atlas, *Chem. Br.*, 1996, *32*(5), 42; (h) J. Kaiser, *Science*, 1999, *284*, 247; (i) T.A. Birkland, *Environment*, 1998, *40*(7), 4; (j) C. Holden, *Science*, 1998, *280*, 1697.

32. Z. Wang, M. Fingas, S. Blenkinsopp, G. Sergy, M. Landriault, L. Sigouin, and P. Lambert, *Environ. Sci. Technol.*, 1998, *32*, 2222.

33. P.J. Squillace, J.S. Zogorski, W.G. Wilber, and C.V. Price, *Environ. Sci. Technol.*, 1996, *30*, 1721.

34. J.E. Reuter, C.F. Harrington, W.R. Cullen, D.A. Bright, and K.J. Reimer, *Environ. Sci. Technol.*, 1998, *32*, 3666.

35. (a) J. Grisham, *Chem. Eng. News*, 1999, May 17, 10; (b) P.M. Morse, *Chem. Eng. News*, 1999, Apr. 12, 26; (c) M. McCoy, *Chem. Eng. News*, 1999, Apr. 5, 9; (d) C. Hogue, *Chem. Eng. News*, Mar. 27, 2000, 6.

36. M. Hamer, *New Sci.*, Sept. 14, 1996, 5.

37. (a) D. Hemenway, *N. Eng. J. Med.*, 1998, *339*, 843; (b) L. Miller, *Wilmington DE News J.*, Aug. 30, 2003, A3.

38. Anon., *Univ. Calif. Berkeley Wellness Lett.*, 1997, *13*(4), 8.

39. Bureau of Transportation Statistics, Washington, DC, Mar. 28, 2003.

40. J.H. Kay, *Asphalt Nation*, Crown, New York, 1997; (b) *Technol. Rev.*, 1997, *100*(5), 53.

41. (a) D. Young. *Alternatives to Sprawl*, Lincoln Institute of Land Policy, Cambridge, MA, 1995; (b) P. Newman and J. Kenworthy, *Sustainability and Cities—Overcoming Automobile Dependence*, Island Press, Washington, DC, 1998; (c) R. Cervero, *The Transit Metropolis—A Global Inquiry*, Island Press, Washington, DC, 1998; (d) J. Pelley, *Environ. Sci. Technol.*, 2000, *34*, 13A.

42. T. Hylton, *Save Our Lands, Save Our Towns: A Plan for Pennsylvania*, RB Books, Harrisburg, PA, 1995.

43. C. Wieman, *Technol. Rev.*, 1996, *99*(4), 48.

44. M. O'Meara, *World Watch*, 1998, *11*(5), 8.

45. (a) R.A. Kerr, *Science*, 1998, *281*, 1128; (b) K.S. Deffeyes, *Hubbert's Peak—The Impending World Oil Shortage*, Princeton University Press, Princeton, 2001.

46. (a) A. Gore, *An Inconvenient Truth: The Planetary Emergency of Global Warming and What We Can Do About It*, Melcher/Rodale, 2006; (b) Anon., *Am. Sci.*, 2007, *95*(3), after p. 282; (c) *Environment*, 2004, *46*(10) [issue on global climate change]; (d) B. Hileman, *Chem. Eng. News*, Oct. 6, 2006, 33; Mar. 21, 2005, 47; (e) C. Hogue, *Chem. Eng. News*, Feb. 2, 2009, 11; (f) C. Hogue, *Chem. Eng. News*, June 22, 2009, 10.

47. (a) D.R. Easterling, B. Horton, P.D. Jones, T.C. Peterson, T.R. Karl, D.E. Parker, M.J. Salinger, V. Razuvayev, N. Plummer, P. Jamason, and C.K. Folland, *Science*, 1997, *277*, 364; (b) K. Hasselman, *Science*, 1997, *276*, 914; (c) R.A. Kerr, *Science*, 1996, *273*, 34; (d) B. Hileman, *Chem. Eng. News*, 1999, Aug. 9, 16; Apr. 24, 2000, 9; (e) R.A. Kerr, *Science*, 2000, *288*, 589; (f) D.R. Easterling, G.A. Meehl, C. Permasan, S.A. Changnon, T.R. Karl, and L.O. Mearns, *Science*, 2000, *289*, 2068; (g) R.T. Watson, ed., *Climate Change 2001 Synthesis Report*, Cambridge, University Press, Cambridge, 2001; (h) D. Hanson, *Chem. Eng. News*, Nov. 26, 2007, 7; (i) Intergovernmental Panel on Climate Change Report, Fourth Assessment, Cambridge University Press, Cambridge, UK, Nov. 17, 2007.

48. R.A. Kerr, *Science*, 2007, *317*, 182.

49. K.R. Briffa, T.J. Osborn, *Science*, 1999, *284*, 926.

50. R.D. Alward, *Science*, 1991, *283*, 229.

51. E.W. Lempinen, *Science*, 2006, *314*, 609.

52. R.A. Kerr, *Science*, 1995, *270*, 1565; 1995, *268*, 802.

53. (a) Anon., *Chem. Eng. News*, 1996, Apr. 22, 29; (b) K.P. Shine and W.T. Sturges, *Science*, 2007, *315*, 1804.

54. (a) K.S. Betts, *Environ. Sci. Technol.*, 1998, *32*, 487A; (b) M. Maiss, C.A.M. Brenninkmeijer, *Environ. Sci. Technol.*, 1998, *32*, 3077.

55. (a) B. Hileman, *Chem. Eng. News*, 1997, May 5, 37; (b) P.E. Kauppi, *Science*, 1995, *270*, 1454.

56. (a) P.S. Zurer, *Chem. Eng. News*, 1995, Mar. 13, 27; (b) R.A. Feely, C.L. Sabine, K. Lee, W. Barelson, J. Kleypass, V.J. Fabry, and F.J. Millero, *Science*, 2004, *305*, 362.
57. J. Uppenbrink, *Science*, 1996, *272*, 1122.
58. J. Kaiser, *Science*, 1997, *278*, 217.
59. Anon., *Chem. Ind. (Lond.)*, 1995, 1001.
60. (a) D.K. Hill, *Science*, 1995, *267*, 1911; (b) S.H. Schneider and T.L. Root, eds, *Wildlife Responses to Climate Change—North American Case Studies*, Island Press, Washington, DC, 2001.
61. (a) V. Loeb, V. Siegel, O. Holn-Hansen, R. Hewitt, W. Fraser, W. Trivelpiece, and S. Trivelpiece, *Nature*, 1997, *387*, 897; (b) *Chem. Eng. News*, *1997*, July 7, 37.
62. (a) H.Q.P. Crick, C. Dudley, D.E. Glue, and D.L. Thomson, *Nature*, 1997, *388*, 526; (b) K.M. Reese, *Chem. Eng. News*, 1997, Sept. 8, 64.
63. B. Hileman, *Chem. Eng. News*, 1995, Nov. 27, 18.
64. R.L. Hotz, *Wilmington Delaware News J.*, 1996, Aug. 29, A13.
65. R. Petkewich, *Chem. Eng. News*, Feb. 23, 2009, 56.
66. (a) R.A. Kerr, *Science*, 1998, *281*, 156; (b) W.S. Broecker, *Science*, 1997, *278*, 1582.
67. (a) A.S. Moffat, *Science*, 1997, *277*, 315; (b) D.H. Janzen, *Science*, 1997, *277*, 883.
68. (a) M.M. Halmann, *Chemical Fixation of Carbon Dioxide—Methods for Recycling Carbon Dioxide into Useful Products*, CRC Press, Boca Raton, FL, 1994; (b) M.M. Halmann and M. Steinberg, *Greenhouse Gas Carbon Dioxide Mitigation: Science and Technology*, Lewis Publishers, Boca Raton, FL, 1999; (c) K.M.K. Yu, I. Curcic, J. Gabriel, and S.C.E. Tsang, *ChemSusChem*, 2008, *1*, 893; (d) S.K. Ritter, *Chem. Eng. News*, Apr. 30, 2007, 11.
69. (a) S. Wilkinson, *Chem. Eng. News*, 1997, Oct. 13, 6; (b) E.E. Benson, C.P. Kubiak, A.J. Sathrum, and J.M. Smieja, *Chem. Soc. Rev.*, 2009, *38*, 89.
70. B. Hileman, *Chem. Eng. News*, 1997, June 30, 30.
71. D. Kreuze, *Technol. Rev.*, 1998, *101*(3), 30.
72. (a) R. Repetto and D. Austin, *The Costs of Climate Protection: A Guide for the Perplexed*, World Resources Institute, Washington, DC, 1997; (b) J.J. Romm, *Cool Companies: How the Best Businesses Boost Profits and Productivity by Cutting Greenhouse Gas Emissions*, Island Press, Washington, DC, 1999.
73. R.A. Kerr, *Science*, 1995, *270*, 731.
74. Anon., *Chem. Eng. News*, 1999, June 21, 24.
75. (a) *Chem. Eng. News*, 1997, June 23, 23; (b) *Environ. Sci. Technol.*, 1997, *31*, 180A.
76. Anon., *Chem. Eng. News*, 1997, Apr. 28, 20.
77. B. Hileman, *Chem. Eng. News*, 1997, Oct. 6, 10.
78. Anon., *Chem. Eng. News*, 1997, June 16, 24.
79. Anon., *Economist*, Nov. 4, 2006, 14.
80. M. Voith, *Chem. Eng. News*, Nov. 17, 2008, 15; (b) E.L. Peterson, M. Beger, and Z.T. Richards, *Science*, 2008, *319*, 1759.
81. Anon., *Modern Plastics*, Dec. 2005, 32.
82. Organization for Economic Cooperation and Development, *Urban Energy Handbook—Good Local Practice*, Paris, 1995.
83. (a) A.H. Rosenfeld, J.J. Romm, H. Akbari, and A.C. Lloyd, *Technol. Rev.*, 1997, *100*(2), 52; (b) H. Akbari, S. Davis, S. Dorsano, J. Huang, and S. Winnett, eds, *Cooling Our Communities—A Guidebook on Tree Planting and Light-Colored Surfacing*, U.S. Environmental Protection Agency, PM-221, 22P-2001, GPO Document 055-000-00371-8, Jan. 1992; (c) G. Moll and S. Ebenreck, *Shading Our Cities: A Resource Guide for Urban and Community Forests*, Island Press, Washington, DC, 1989; (d) Green Living Staff, A.S. Moffat, and M. Schiler, *Energy-Efficient and Environmental Landscaping*, Appropriate Solutions Press, South Newfane, VT, 1994.
84. A. Meier, ed., *Energy Buildings*, 1997, *25*, 99–177.

85. S.E. Bretz and H. Akbari, *Energy Buildings*, 1997, *25*, 159.

86. A. Heller, *Acc. Chem. Res.*, 1995, *28*, 503.

87. (a) J.D. Balcomb, *Passive Solar Buildings*, MIT Press, Cambridge, MA., 1992; (b) P. O'Sullivan, *Passive Solar Energy in Buildings*, Elsevier Applied Science, New York, 1988; (c) C. Carter and J. de Villiere, *Principles of Passive Solar Building Design*, Pergamon, New York, 1987; (d) A. Bisio and S.R. Boots, *Energy Technology and the Environment*, Wiley, New York, 1995, 468; (e) Union of Concerned Scientists, *Put Renewable Energy to Work in Buildings*, Cambridge, MA, Jan. 1993; (f) J. Nieminen, *Energy Buildings*, 1994, *21*, 187; (g) N. Lenssen and D.M. Roodman, *State of the World 1995*, Worldwatch Institute, W.W. Norton, New York, 1995, 95; (h) F. Sick and J. Leppanen, *Sol. Energy*, 1994, *53*, 379; (i) Steven Winter Associates, *The Passive Solar Design and Construction Handbook*, Wiley, New York, 1997; (j) www.ucsusa.org/clean_energy; (k) www.eere/energy.gov/RE/passive_passive.html; (l) www.nrdc.org/buildinggreen/strategies/energy.asp; (m) Clinton Foundation, "Energy Efficiency Building Retrofit Program," Fact Sheet, May 2007, New York; (n) A.K. Athienitis and M. Santamouris, *Thermal Analysis and Design of Passive Solar Buildings*, Stylus Publishing, Herndon, VA, 2002.

88. C. Filippin, A. Beascochea, A. Esteves, C. de Rosa, L. Cortegoso, and D. Estelrich, *Sol. Energy*, 1998, *63*, 105.

89. (a) G. Rockendorf, S. Janssen, and H. Felten, *Sol. Energy*, 1996, *58*, 33; (b) H. Gunnewick, E. Brundrett, and K.G.T. Hollands, *Sol. Energy*, 1996, *58*, 227.

90. G. Oliveti and N. Arcuri, *Sol. Energy*, 1996, *57*, 345.

91. S.N.G. Lo and B. Norton, *Sol. Energy*, 1996, *56*, 143.

92. M. Ronnelid and B. Karlsson, *Sol. Energy*, 1996, *57*, 93.

93. P. Kariukinyahoro, R.R. Johnson, and J. Edwards, *Sol. Energy*, 1997, *59*, 11.

94. A. Balzar, P. Stumpf, S. Eckhoff, H. Ackermann, and M. Grupp, *Sol. Energy*, 1996, *58*, 63.

95. (a) M. Reuss, *Sol. Energy*, 1997, *59*, 259; (b) R.J. Fuller and W.W.S. Charters, *Sol. Energy*, 1997, *59*, 151; (c) C. Palaniappan and S.V. Subramanian, *Sol. Energy*, 1998, *63*, 31.

96. R. Croy and F.A. Peuser, *Sol. Energy*, 1994, *53*(1), 47.

97. (a) N. Chege, *World Watch*, 1995, *8*(5), 8; (b) J.D. Burch and K.E. Thomas, *Sol. Energy*, 1998, *64*, 87.

98. (a) G. Mink, M.M. Aboabboud, and E. Karmazsin, *Sol. Energy*, 1998, *62*, 309; (b) www.solar-desalination.com/4.html.

99. G.S. Samdani, *Chem. Eng.*, 1995, *102*(3), 19.

100. (a) M. Santamouris, C.A. Balaras, E. Dascalaki, and M. Vallindras, *Sol. Energy*, 1994, *53*, 411; (b) I.A. Abbud, G.O.G. Lof, and D.C. Hittle, *Sol. Energy*, 1995, *54*, 75.

101. (a) K. Morino and T. Oka, *Energy and Buildings*, 1994, *21*, 65; (b) M. Reuss, M. Beck, and J.P. Muller, *Sol. Energy*, 1997, *59*, 247; (c) D.S. Breger, J.E. Hubbell, H. El Hasnaqui, and J.E. Sunderland, *Sol. Energy*, 1996, *56*, 493.

102. (a) R. Kubler, N. Fisch, and E. Hahne, *Sol. Energy*, 1997, *61*, 97; (b) G. Oliveti and N. Arcuri, *Sol. Energy*, 1995, *54*, 85.

103. M.T. Kangas and P.D. Lund, *Sol. Energy*, 1994, *53*, 237.

104. G. Oliveti, N. Arcuri, and S. Ruffolo, *Sol. Energy*, 1998, *62*, 281.

105. (a) P. Brousseau and M. Lacroix, *Energy Convers. Manage.*, 1996, *37*, 599; (b) D.W. Hawes and D. Feldman, *Sol. Energy Mater. Sol. Cells*, 1992, *27*, 91, 103.

106. G.A. Lane, *Sol. Energy Mater. Sol. Cells*, 1992, *27*, 135.

107. H.W. Ryu, S.W. Woo, B.C. Shin, and S.D. Kim, *Sol. Energy Mater. Sol. Cells*, 1992, *27*, 161.

108. (a) A.H. Rosenfeld and D. Hafemeister, *Sci. Am.*, 1988, Apr. 78; (b) J. Johnson, *Chem. Eng. News*, Dec. 3, 2007, 46.

109. N. Henbest, *New Sci.*, 1989, Feb. 11, 44.

110. G. Weaver, private communication.

111. Anon., *Chem. Eng. News*, Apr. 22, 2002, 56.

112. (a) Anon., *Environment*, 1994, *36*(10), 24; (b) *Chem. Ind. (Lond.)*, 1994, 844.

113. T. Justel, H. Nikol, and C. Ronda, *Angew. Chem. Int. Ed.*, 1998, *37*, 3084.

114. (a) M. Rouhi, *Chem. Eng. News*, 1999, Feb. 8, 37; (b) R.T. Wegh, H. Donker, K.D. Oskam, and A. Meijerink, *Science*, 1999, *283*, 663.

115. Cermax Lamp Engineering Guide ILC Technology, Sunnyvale, CA, 1998.

116. Anon., *R&D (Cahners)* 1998, *40*(4), 100.

117. (a) Anon., *R&D (Cahners)*, 1998, *40*(4), 7; (b) B. Gill, ed., *Group III Nitride Semiconductor Compounds: Physics and Applications*, Oxford University Press, New York, 1998; (c) D. Wilson, *Technol. Rev.*, 2000, *103*(3), 27; (d) Y.S.L. Chung, M.Y. Jeon, and C.K. Kim, *Ind. Eng. Chem. Res.*, 2009, 48, 740; (e) Anon., *Technol. Rev.*, 2009, 112(4), 16; (f) H.A. Hoppe, *Angew. Chem. Ind. Ed.*, 2009, 48, 3572.

118. (a) J. Walker, *R&D (Cahners)*, 2004, *46*(5), 56; (b) D. Talbot, *Technol. Rev.*, 2003, *106*(4), 30; (c) J. Johnson, *Chem. Eng. News*, Dec. 3, 2007, 46.

119. (a) J.D. Anderson, E.M. McDonald, P.A. Lee, M.L. Anderson, E.L. Ritchie, H.K. Hall, T. Hopkins, E.A. Mash, J. Wang, A. Padias, S. Thayumanavan, S. Barlow, S.R. Marder, G.E. Jabbour, S. Shaheen, B. Kippelen, N. Peyghambarian, R.M. Wightman, and N.R. Armstrong, *J. Am. Chem. Soc.*, 1998, *120*, 9646; (b) W.-L. Yu, J. Pei, Y. Cao, W. Huang, and A.J. Heeger, *Chem. Commun.*, 1999, 1837; (c) P. Livingstone, *R&D (Cahners)*, 2008, *50*(1), 43; (b) K. Mullen and U. Scherf, *Organic Light Emitting Devices: Synthesis, Properties and Applications*, Wiley-VCH, Weinheim, 2005; (d) E. Polikparov and M.E. Thompson, *Mater. Matters (Aldrich)*, Milwaukee, WI, 2007, *3*(2), 21.

120. (a) A. Johansson, *Clean Technology*, Lewis Publishers, Boca Raton, FL, 1992, 74–94; (b) T.B. Johansson, H. Kelly, A.K.N. Reddy, and R.H. Williams, eds, *Renewable Energy—Sources for Fuels and Electricity*, Island Press, Washington, DC, 1993; (c) E.A. Torrero, *Kirk–Othmer Encyclopedia of Chemical Technology*, 4th ed., New York, 1997, *21*, 218–236; (d) J.A.G. Drake, ed., *Electrochemistry and Clean Energy*, Royal Society of Chemistry, Cambridge, 1994; (e) K.S. Brown, *Science*, 1999, *285*, 678; (f) V. Comello, *R&D (Cahners)*, 1998, *40*(5), 22; (g) L. Freris and D. Infield, *Renewable Energy in Power Systems*, Wiley, New York, 2008; (h) EUREC Agency, *The Future of Renewable Energy 2, Prospects and Directions*, Stylus Publishing, Herndon, VA, 2002; (i) D. Assmann, U. Laumanns, and D. Uh, eds, *Renewable Energy—A Global Review of Technologies, Policies and Markets*, Stylus Publishing, Herndon, VA, 2006.

121. (a) F.S. Sterrett, ed., *Alternate Fuels and the Environment*, Lewis Publishers, Boca Raton, FL, 1994; (b) W.M. Kreucher, *Chem. Ind. (Lond.)*, 1995, 601; (c) D.C. Carslaw and N. Fricker, *Chem. Ind. (Lond.)*, 1995, 593; (d) Union of Concerned Scientists, *Alternative Transportation Fuels*, Cambridge, MA, Dec. 1991.

122. (a) D. Sperling, *Chem. Ind. (Lond.)*, 1995, 609; (b) J.J. MacKenzie, *The Keys to the Car: Electric and Hydrogen Vehicles for the 21st Century*, World Resources Institute, Washington, DC, 1994.

123. L.B. Lave, C.T. Handrickson, and F.C. McMichael, *Science*, 1995, *268*, 993.

124. K. Sasaki, M. Yokota, H. Nagayoshi, and K. Kamisako, *Sol. Energy Mater. Sol. Cells*, 1997, *47*, 259.

125. (a) A.M. Rouhi, *Chem. Eng. News*, 1995, May 29, 37; (b) S. Samdani, *Chem. Eng.*, 1995, *102*(4), 17; (c) R.W. Wegman, *J. Chem. Soc. Chem. Commun.*, 1994, 947; (d) Q. Ge, Y. Huang, F. Qiu, and S. Li, *Appl. Catal. A*, 1998, *167*, 23; (e) G. Parkinson, *Chem. Eng.*, 1999, *106*(10), 19.

126. (a) Anon., *Environ. Sci. Technol.*, 1965, *29*, 67A; (b) *Environment*, 1998, *40*(1), 22.

127. C.A. Grinder, *Waste News*, 1999, *5*(12), 1.

128. M. Murray, *Wilmington Delaware News J.*, 1995 July 12, B1.
129. (a) K. Fouhy and S. Shelley, *Chem. Eng.*, 1997, *104*(5), 55; (b) C. He, D.J. Herman, R.G. Minet, and T.-T. Tsotsis, *Ind. Eng. Chem. Res.*, 1997, *36*, 4100; (c) B. Eklund, E.P. Anderson, B.L. Walker, and D.B. Burrows, *Environ. Sci. Technol.*, 1998, *32*, 2233.
130. (a) A. Kendall, A. McDonald, and A. Williams, *Chem. Ind.* (*Lond.*), 1997 342; (b) D.P.L. Murphy, ed., *Energy from Crops*, Semundo, Cambridge 1996; (c) V.R.Tolbert and A. Schiller, *Am. J. Alt. Agric.*, 1996, *11*, 148–149 (d) W. Patterson, *Power from Plants: The Global Implications of New Technologie, for Electricity from Biomass*, Earthscan, London, 1994; (e) M.R. Khan, ed. *Clean Energy from Waste and Coal*, A.C.S. Symp. 515, Washington, DC, 1992 (f) T.B. Johansson, H. Kelly, A.K.N. Reddy, and R.H. Williams, eds, *Renewabl, Energy: Sources for Fuels and Electricity*, Island Press, Washington, DC, 1993, *38* 56; (g) A. Faaij, *Energy from Biomass and Waste*, Utrecht University, Utrecht, 1997 (h) A.V. Bridgwater and D.G.B.L. Boocock, eds, *Developments in Thermochemica, Biomass Conversion*, Chapman & Hall, London, 1997; (i) T.E. Bull, *Science*, 1999 *285*, 1209; (j) J.A. Turner, *Science*, 1999, *285*, 1209.
131. Anon., *Chem. Eng. News*, 1999, Aug. 23, 30.
132. D.M. Kammen, *Environment*, 1999, *41*(5), 10.
133. Anon., *Environment*, 1999, *41*(1), 20.
134. F. Pearce, *New Sci.*, 1995, Jan. 14, 12.
135. (a) G.S. Samdani, *Chem. Eng.*, 1994, *101*(11), 19; (b) G. Parkinson, *Chem. Eng.*, 199 *104*(10), 23.
136. (a) A.V. Bridgwater, *Appl. Catal.*, 1994, *116*, 5; (b) C. Zhao, Y. Kou, A.A. Lemonido, X. Li, and J.A. Lercher, *Angew. Chem. Int. Ed.*, 2009, *48*, 3987.
137. (a) J.M. Ogden and R.H. Williams, *Solar Hydrogen—Moving Beyond Fossil Fuel* World Resources Institute, Washington, DC, 1989; (b) J.C. Cannon, Harnessir Hydrogen: The Key to Sustainable Transportation, *Int. News Fats, Oils Relat. Mate,* New York, 1995; (c) T.B. Johansson, H. Kelley, A.K.N. Reddy, and R.H. Williams, ed *Renewable Energy: Sources for Fuels and Electricity*, Island Press, Washington, D(1993, 925; (d) D.A.J. Rand and R.M. Dell, Hydrogen economy: Challenges and pro pects, *Royal Soc. Chem. Publishing*, Cambridge, UK, 2007.
138. W. Lepkowski, *Chem. Eng. News*, 1995, Mar. 6, 22; 1996, Oct. 21, 31.
139. (a) K.V. Kordesch and G.R. Simader, *Chem. Rev.*, 1995, *95*, 191; (b) J.D. Holladay, J. H D.L. King, and Y. Wang, *Catal. Today*, 2009, *139*, 244.
140. V.N. Parmon, *Catal. Today*, 1997, *35*, 153.
141. (a) S.E. Nilsen and K. Andreassen, *Int. News Fats, Oils Relat. Mater.*, 1996, *7*, 11, (b) B.E. Logan, *Environ. Sci. Technol.*, 2004, *38*, 160A.
142. (a) T. Takata, A. Tanaka, M. Hara, J.N. Kondo, and K. Domen, *Catal. Today*, 1998, , 17; (b) J.R. Durrant, *Chem. Ind.* (*Lond.*), 1998, 838; (c) J. Jacoby, *Chem. Eng. Ne,* August 10, 2009, 7.
143. (a) A. Heller, *Acc. Chem. Res.*, 1995, *28*, 503; (b) J.R. Bolton, *Sol. Energy*, 1996, *57*, (c) Anon., *Chem. Eng. News*, Oct. 2, 2000, 45.
144. (a) O. Khaselev and J.A. Turner, *Science*, 1998, *280*, 425; (b) S.S. Kocha, D. Montgom(M.W. Peterson, and J.A. Turner, *Sol. Energy Mater. Sol. Cells*, 1998, *52*, 389.
145. (a) K. Sayama, H. Arakawa, and K. Domen, *Catal. Today*, 1996, *28*, 175; (b) J. Yoshim, A. Tanaka, J.N. Kondo, and K. Domen, *Bull. Chem. Soc. Jpn.*, 1995, *68*, 24 (c) M. Freemantle, *Chem. Eng. News*, 1999, June 28, 10; (d) H.G. Kim, D.W. Hw, J. Kim, Y.G. Kim, and J.S. Lee, *Chem. Commun.*, 1999, 1077; (e) W. Shanggt K. Inoue, and A. Yoshida, *Chem. Commun.*, 1998, 779; (f) D.W. Hwang, H.G. K J. Kim, K.Y. Cho, Y.G. Kim, and J.S. Lee, *J. Catal.*, 2000, *193*, 40.
146. K.-H. Chung and D.-C.Park, *Catal. Today*, 1996, *30*, 157.

102   Problem-Solving Exercises in Green and Sustainable Chemistry

147. (a) S. Ikeda, T. Takata, T. Kondo, G. Hitoki, M. Hara, J.N. Kondo, K. Domen, H. Hosono, H. Kawazoe, and A. Tanaka, *Chem. Commun.*, 1998, 2185; (b) M. Hara, T. Kondo, M. Komoda, S. Ikeda, K. Shinohara, A. Tanaka, J.N. Kondo, and K. Domen, *Chem. Commun.*, 1998, 357.

148. S. Ichikawa and R. Doi, *Catal. Today*, 1996, *27*, 271.

149. K. Tennakone, G.R.R.A. Kumara, I.R.M. Kottegoda, and V.P.S. Perera, *Chem. Commun.*, 1999, 15.

150. G. Parkinson, *Chem. Eng.*, 1998, *105*(10), 21.

151. (a) J. Graetz, *Chem. Soc. Rev.*, 2009, *38*, 73; (b) A.W.C. van den Berg and C.O. Arean, *Chem. Commun.*, 2008, 668; (c) M. Jacoby, *Chem. Eng. News*, Jan. 28, 2008, 67; (d) U. Eberle, M. Felderhoff, and F. Schuth, *Angew. Chem. Int. Ed.*, 2009, 48, 6608.

152. C. Read, G. Thomas, G. Ordaz, and S. Satyapal, *Mater. Matters (Aldrich)*, Milwaukee, WI, 2007, *2*(2), 3.

153. (a) A. Karkambar, C. Ardahl, T. Autrey, and G.L. Soloveichik, *Mater. Matters (Aldrich)*, Milwaukee, WI, 2007, *2*(2), 6, 11; (b) T.B. Marder, *Angew. Chem. Int. Ed.*, 2007, *46*, 8116.

154. (a) Y.H. Hu and E. Ruckenstein, *Ind. Eng. Chem. Res.*, 2008, *47*, 48; (b) J. Yang, A. Sudik, D.J. Siegel, D. Halliday, A. Drews, R.O. Carter, III, C. Wolverton, G.J. Lewis, J.W.A. Sachtler, J.J. Low, S.A. Faheem, D.A. Lesch, and V. Ozolins, *Angew. Chem. Int. Ed.*, 2008, *47*, 882.

155. (a) J.L.C. Rowsell and O.M. Yaghi, *Angew. Chem. Int. Ed.*, 2005, *44*, 4670; (b) M. Jacoby, *Chem. Eng. News*, May 19, 2003, 11.

156. L. Schlapbach and A. Zuttel, *Nature*, 2001, *414*, 353.

157. (a) E.D. Naeemi, D. Graham, and B.F. Norton, *Mater. Matters (Aldrich)*, Milwaukee, WI, 2007, *2*(2), 23; (b) N. Kariya, A. Fukuoka, and M. Ichikawa, *Chem. Commun.*, 2003, 690; (c) P. Ferreira-Aparicio, I. Rodriguez-Ramos, and A. Guerrero-Ruoz, *Chem. Commun.*, 2002, 2082; (d) G. Parkinson, *Chem. Eng.*, 2003, *110*(3), 21.

158. (a) W. Vielstich, H. Gasteiger, and A. Lamm, *Handbook of Fuel Cells—Fundamentals, Technology and Applications*, Wiley-VCH, Weinheim, 2003, *4*; (b) F. de Bruijn, *Green Chem.*, 2005, *7*(3), 132; (c) E. Dorey, *Chem. Ind. (Lond.)*, Feb. 12, 2007, 10; (d) H.W. Cooper, *Chem. Eng. Prog.*, 2007, *103*(11), 14; (e) R. Datta, *Fuel Cell Principles, Components and Assemblies*, Wiley-Blackwell, Hoboken, NJ, 2009.

159. E.J. Hoffman, *Power Cycles and Energy Efficiency*, Academic, San Diego, 1996.

160. (a) M. Miller, *Hydrogen Fuel Cell Vehicles*, Union of Concerned Scientists, Cambridge, MA, Feb. 1995; (b) T. Newton, *Chem. Br.*, 1997, *33*(1), 29; (c) K. Fouhy and G. Ondrey, 1996, *103*(8), 46; (d) R. Edwards, *New Sci.*, 1996, *152*(2057), 40; (e) *Chem. Eng. News*, 1996, May 20, 38; (f) T.R. Ralph and G.A. Hards, *Chem. Ind. (Lond.)*, 1998, 337; (g) V. Raman, *Chem. Ind. (Lond.)*, 1997, 771; (h) J. Yamaguchi, *Automot. Eng.*, 1998, *106*(7), 54.

161. Anon., *Appl. Catal. B*, 1997, *12*, N14.

162. (a) Anon., *Chem. Eng. News*, 1997, Oct. 27; (b) B. Hileman, *Chem. Eng. News*, 1997, Apr. 21, 31; (c) Anon., *Chem. Eng. News*, 1997, Jan. 13, 20; (d) Anon., *Chem. Eng. News*, 1997, Aug. 4, 16; (e) A. Hamnett, *Catal. Today*, 1997, *38*, 445; (f) C. Sloboda, *Chem. Eng.*, 1997, *104*(11), 49; (g) M. Jacoby, *Chem. Eng. News*, 1999, Aug. 16, 7; (h) Anon., *Chem. Eng. News*, 1999, Jan. 11, 22; (i) S. Velu, K. Suzuki, and T. Osaki, *Chem. Commun.*, 1999, 2341; (j) R. Corfield, *Chem. Ind. (Lond.)*, Aug. 21, 2006, 22.

163. (a) S. Park, J.M. Vohs, and R.J. Gorte, *Nature*, 2000, *404*, 265; (b) T. Hibino, A. Hashimoto, T. Inoue, J.-I. Tokumo, S.-I.Yoshida, and M. Sano, *Science*, 2000, *288*, 2031.

164. Y. Hishinuma and M. Kunikata, *Energy Convers. Manage.*, 1997, *38*, 1237.

165. (a) Anon., *Chem. Ind. (Lond.)*, 1996, 823; (b) D. Hart, *Chem. Ind. (Lond.)*, 1998, 344; (c) V. Thangadurai, A.K. Shukla, and J. Gopalakrishnan, *Chem. Commun.*, 1998, 2647; (d) D.J.L. Brett, A. Atkinson, N.P. Brandon, and S.J. Skinner, *Chem. Soc. Rev.*, 2008,

37, 1568; (e) S. Park, J.M. Vohs, and R.J. Gorte, *Nature*, 2000, *404*, 265; (f) L. Yang, S. Wang, K. Blinn, M. Liu, Z. Liu, Z. Cheng, and M. Liu, *Science*, 2009, *326*, 126; (g) T. Suzuki, Z. Hasan, Y. Funahashi, T. Yamaguchi, Y. Fujishiro, and M. Awano, *Science*, 2009, *325*, 126.

166. K.R. Williams and G.T. Burstein, *Catal. Today*, 1997, *38*, 401.

167. J.J. Romm, *The Hype about Hydrogen: Fact and Fiction in the Race to Save the Planet*, Island Press, Washington, DC.

168. (a) D.A. Beattie, *History and Overview of Solar Heat Technologies*, MIT Press, Cambridge, MA, 1997; (b) S. Moran and J.T. McKinnon, *World Watch*, 2008, *21*(2), 26; (c) Anon., *Technol. Rev.*, 2007, *110*(5), S1; (d) R. Peltier, *Power*, 2007, *151*(12), 40; (e) E. Corcoran, *Forbes*, Sept. 3, 2007, 80–92; (f) D. Rotman, *Technol. Rev.*, 2009, *112*(4), 47; (g) G. Ondrey, *Chem. Eng.*, 2009, *116*(2), 11; (h) G. Johnson, *Nat. Geo. Mag.*, 2009, *215*(9), 28.

169. Q.-C. Zhang, Y. Yin, and D.R. Mills, *Sol. Energy Mater. Sol. Cells*, 1996, *40*, 43.

170. T. Eisenhammer, *Sol. Energy Mater. Sol. Cells*, 1997, *46*, 53.

171. Y. Yin, D.R. McKenzie, and W.D. McFall, *Sol. Energy Mater. Sol. Cells*, 1996, *44*, 69.

172. Q.-C. Zhang, K. Zhao, B.-C.Zhang, L.-F.Wang, Z.-L.Shen, Z.-J.Zhou, D.-L.Xie, and B.-F. Li, *Sol. Energy*, 1998, *64*, 109.

173. E. Barrera, I. Gonzalez, and T. Viveros, *Sol. Energy Mater. Sol. Cells*, 1998, *51*, 69.

174. (a) D.Y. Coswami, ed., *Advances in Solar Energy*, Stylus Publishing, Herndon, VA, 2005 [an annual publication]; (b) H. Scheer, *The Solar Economy*, Stylus Publishing, Herndon, VA, 2004; (c) German Solar Energy Society, *Planning and Installing Photovoltaic Systems, A Guide for Installers, Architects and Engineers*, Stylus Publishing, Herndon, VA, 2005; (d) R. Eisenberg, ed., *Inorg. Chem.*, 2005, *44*, 6799–6911; (e) A. Luque and S. Hegedus, eds, *Handbook of Photovoltaic Science and Engineering*, Wiley, New York, 2003; (f) A.H. Tullo, *Chem. Eng. News*, Nov. 20, 2006, 25; (g) J. Johnson, *Chem. Eng. News*, Oct. 20, 2008, 40; (h) M. Behar, *On Earth (Natural Resources Defense Council)*, 2009, *31*(1), 26.

175. (a) Anon., *Chem. Eng. News*, 1997, July 7, 30; (b) J. Johnson, *Chem. Eng. News*, 1998, Mar. 30, 24.

176. P.D. Maycock, *P.V. News*, Sept.–Oct., 18.

177. (a) M. Rynn, *Amicus J. (Natural Resources Defense Council)*, 1999, *21*(2), 15; (b) C. Flavin and M. O'Meara, *World Watch*, 1998, *11*(5), 23; (c) J. Johnson, *Chem. Eng. News*, Oct. 9, 2000, 45; (d) R.F. Service, *Science*, 2008, *319*, 718.

178. (a) A.S. Bouazzi, M. Abaab, and B. Rezig, *Sol. Energy Mater. Sol. Cells*, 1997, *46*, 29; (b) J. Zhao, A. Wang, P. Altermat, S.R. Wenham, and M.A. Green, *Sol. Energy Mater. Sol. Cells*, 1996, *41/42*, 87.

179. W. Wettling, *Sol. Energy Mater. Sol. Cells*, 1995, *38*, 487, 494.

180. (a) A. Hubner, C. Hampe, and A.G. Aberle, *Sol. Energy Mater. Sol. Cells*, 1997, *46*, 67; (b) M. Ghannam, S. Sivoththaman, J. Poortmans, J. Szlufcik, J. Nijs, R. Mertens, and R. van Overstraeten, *Sol. Energy*, 1997, *59*, 101; (c) M. Pauli, T. Reindl, W. Kruhler, F. Holmberg, and J. Muller, *Sol. Energy Mater. Sol. Cells*, 1996, *41/42*, 119; (d) A.U. Ebong, C.B. Honsberg, and S.R. Wenham, *Sol. Energy Mater. Sol. Cells*, 1996, *44*, 271; (e) F.C. Marques, J. Urdanivia, and I. Chambouleyron, *Sol. Energy Mater. Sol. Cells*, 1998, *52*, 285.

181. P. Doshi, A. Rohatgi, M. Ropp, Z. Chen, D. Ruby, and D.L. Meier, *Sol. Energy Mater. Sol. Cells*, 1996, *41/42*, 31.

182. L.C. Klein, ed., *Sol–Gel Technology*, Noyes, Park Ridge, NJ, 1988, 80.

183. Y. Yazawa, T. Kitatani, J. Minemura, K. Tamura, K. Mochizuki, and T. Warabisako, *Sol. Energy Mater. Sol. Cells*, 1994, *35*, 39.

184. F.Z. Tepehan, F.E. Ghodsi, N. Ozer, and G.G. Tepehan, *Sol. Energy Mater. Sol. Cells*, 1997, *46*, 311.

185. E. Aperathitis, Z. Hatzopoulos, M. Androulidaki, V. Foukaraki, A. Kondilis, C.G.Scott, D. Sands, and P. Panayotatos, *Sol. Energy Mater. Sol. Cells*, 1997, *45*, 161.

186. (a) F. Zhu, T. Fuyuki, H. Matsunami, and J. Singh, *Sol. Energy Mater. Sol. Cells*, 1995, *39*, 1; (b) M. Martinez, J. Herrero, and M.T. Gutierrez, *Sol. Energy Mater. Sol. Cells*, 1997, *45*, 75.

187. Anon., *Chem. Eng. News*, Mar. 5, 2007, 52.

188. S. Walheim, E. Schaffer, J. Mlynek, and U. Steiner, *Science*, 1999, *283*, 520.

189. S. Winderbaum, O. Reinhold, and F. Yun, *Sol. Energy Mater. Sol. Cells*, 1997, *46*, 239.

190. (a) P. Faith, C. Borst, C. Zechner, E. Bucher, G. Willeki, and S. Narayana, *Sol. Energy Mater. Sol. Cells*, 1997, *48*, 229; (b) T. Machida, A. Miyazawa, Y. Yokosawa, H. Nakaya, S. Tanaka, T. Nunoi, H. Kumada, M. Murakami, and T. Tomita, *Sol. Energy Mater. Sol. Cells*, 1997, *48*, 243.

191. (a) F.J. Pern, *Sol. Energy Mater. Sol. Cells*, 1996, *41/42*, 587; (b) A.W. Czanderna and F.J. Pern, *Sol. Energy Mater. Sol. Cells*, 1996, *43*, 101; (c) P. Klemchuk, M. Ezrun, G. Lavigne, W. Holley, J. Galica, and S. Agro, *Polym. Degrad. Stabil.*, 1997, *55*, 347; (d) F.J. Pern, *Angew. Makromol. Chem.*, 1997, *252*, 195.

192. P. Walter, *Chem. Ind. (Lond.)*, Dec. 18, 2006, 11.

193. (a) A. Rohatgi and S. Narasimha, *Sol. Energy Mater. Sol. Cells*, 1997, *48*, 187; (b) J. Nijs, S. Sivoththaman, J. Szlufcik, K. De Clercq, F. Duerinckx, E. Van Kerschaever, R.Einhaus, J. Poortmans, T. Vermeulen, and R. Mertens, *Sol. Energy Mater. Sol. Cells*, 1997, *48*, 199; (c) Z. Yuwen, L. Zhongming, M. Chundong, H. Shaoqi, L. Zhiming, Y. Yuan, and C. Zhiyun, *Sol. Energy Mater. Sol. Cells*, 1997, *48*, 167; (d) A.V. Shah, R. Platz, and H. Keppner, *Sol. Energy Mater. Sol. Cells*, 1995, *38*, 501; (e) M. Reisch, *Chem. Eng. News*, July 30, 2007, 15.

194. G. Ondrey, *Chem. Eng.*, 2004, *111*(13), 15; 2007, *114*(8), 14.

195. (a) A.V. Shah, R. Platz, and H. Keppner, *Sol. Energy Mater. Sol. Cells*, 1995, *38*, 505; (b) E.A. Schiff, A. Matsuda, M. Hack, S.J. Wagner, and R. Schropp, eds, *Amorphous Silicon Technology: 1995, 1996, and 1997*, Materials Research Society, Pittsburgh, PA, 1995–1997; (c) C. Eberspacher, H.W. Schock, D.S. Ginley, T. Catalano, and T. Wada, eds, *Thin Films for Photovoltaic and Related Device Applications*, Materials Research Society, Pittsburgh, PA, 1996, *426*; (d) B. Jagannathan, W.A. Anderson, and J. Coleman, *Sol. Energy Mater. Sol. Cells*, 1997, *46*, 289.

196. R.F. Service, *Science*, 1998, *279*, 1300.

197. (a) F. Demichelis, G. Crovini, C.F. Pirri, E. Tresso, R. Galloni, R. Rizzoli, C.Summonte, and P. Rava, *Sol. Energy Mater. Sol. Cells*, 1995, *37*, 315; (b) H. Nishiwaki, K. Uchihashi, K. Takaoka, M. Nakagawa, H. Inoue, A. Takeoka, S. Tsuda, and M. Ohnishi, *Sol. Energy Mater. Sol. Cells*, 1995, *37*, 295; (c) C.R. Wronski, *Sol. Energy Mater. Sol. Cells*, 1996, *41/42*, 427; (d) Y. Hishikawa, K. Ninomiya, E. Maruyama, S. Kuroda, A. Terakawa, K. Sayama, H. Tarui, M. Sasaki, S. Tsuda, and S. Nakano, *Sol. Energy Mater. Sol. Cells*, 1996, *41/42*, 441; (e) B. Rech, C. Beneking, and H. Wagner, *Sol. Energy Mater. Sol. Cells*, 1996, *41/42*, 475; (f) M. Kubon, E. Boehmer, F. Siebke, B. Rech, C. Beneking, and H. Wagner, *Sol. Energy Mater. Sol. Cells*, 1996, *41/42*, 485; (g) M.S. Haque, H.A. Naseem, and W.D. Brown, *Sol. Energy Mater. Sol. Cells*, 1996, *41/42*, 543; (h) N. Gonzalez, J.J. Gandia, J. Carabe, and M.T. Gutierrez, *Sol. Energy Mater. Sol. Cells*, 1997, *45*, 175; (i) R. Martins, *Sol. Energy Mater. Sol. Cells*, 1997, *45*, 1.

198. A. Shah, P. Torres, R. Tscharner, N. Wyrsch, and H. Keppner, *Science*, 1999, *285*, 692.

199. K. Winz, C.M. Fortman, T. Eickhoff, C. Beneking, H. Wagner, H. Fujiwara, and I. Shimizu, *Sol. Energy Mater. Sol. Cells*, 1997, *49*, 195.

200. (a) T. Yamawaki, S. Mizukami, A. Yamazaki, and H. Takahashi, *Sol. Energy Mater. Sol. Cells*, 1997, *47*, 125; (b) M. Kondo, H. Nishio, S. Kurata, K. Hayashi, A. Takenaka, A. Ishikawa, K. Nishimura, H. Yamagishi, and Y. Tawada, *Sol. Energy Mater. Sol. Cells*, 1997, *49*, 1.

201. S. Guha, J. Yang, A. Banerjee, T. Glatfelter, and S. Sugiyama, *Sol. Energy Mater. Sol. Cells*, 1997, *48*, 365.

202. T. Yoshida, S. Fujikake, S. Kato, M. Tanda, K. Tabuchi, A. Takano, Y. Ichikawa, and H. Sakai, *Sol. Energy Mater. Sol. Cells*, 1997, *48*, 383.

203. (a) F. Roca, G. Sinno, G. DiFrancia, P. Prosini, G. Pascarella, and D. Della Sala, *Sol. Energy Mater. Sol. Cells*, 1997, *48*, 15; (b) G. Ballhorn, K.J. Weber, S. Armand, M.J. Stocks, and A.W. Blakers, *Sol. Energy Mater. Sol. Cells*, 1998, *52*, 61.

204. E. Kintisch, *Science*, 2007, *317*, 583.

205. (a) K.S. Betts, *Envion. Sci. Technol.*, 2007, *41*, 3038; (b) Anon., *Chem. Eng. Prog.*, 2008, *104*(9), 18.

206. (a) P. Pechy, F.P. Rotzinger, M.K. Nazeeruddin, O. Kohle, S.M. Zakeeruddin, R. Humphry-Baker, and M. Gratzel, *J. Chem. Soc. Chem. Commun.*, 1995, 65; (b) A. Coghlan, *New Sci.*, 1995, Jan. 28, 22; (c) M. Freemantle, *Chem. Eng. News*, 1998, Oct. 26, 44; (d) M.K. Nazeeruddin, R. Humphry-Baker, M. Gratzel, and B.A. Murrer, *Chem. Commun.*, 1998, 719; (e) H. Sugihara, L.P. Singh, K. Seyama, H. Arakawa, M.K. Nazeeruddin, and M. Gratzel, *Chem. Lett.*, 1998, *27*, 1005; (f) U. Bach, D. Lupo, P. Comte, J.E. Moser, F. Weissortel, J. Salbeck, H. Sprietzer, and M. Gratzel, *Nature*, 1998, *395*, 583; (g) W.C. Sinke and M.M. Wienk, *Nature*, 1998, *395*, 544; (h) A. Hagfeldt and M. Gratzel, *Acc. Chem. Res.* 2000, *33*, 269; (i) N.S. Lewis, *ChemSusChem*, 2009, *2*, 383.

207. (a) N. Robertson, *Angew. Chem. Int. Ed.*, 2006, *45*, 2338; (b) J.-H.Yum, D.P. Hagberg, S.-J. Moon, K.M. Karlson, T. Marinado, L. Sun, A. Hagfeldt, M.K. Nazeeruddin, and M. Gratzel, *Angew. Chem. Int. Ed.*, 2009, *48*, 1576; (c) A. Mishra, M.K.L.R. Fischer and P. Baurele, *Angew. Chem. Int. Ed.*, 2009, *48*, 2474.

208. R.F. Service, *Science*, 1996, *272*, 1744.

209. Anon., *Environment*, 1999, *41*(5), 22.

210. C. Flavin and M. O'Meara, *World Watch*, 1997, *10*(3), 28.

211. (a) J.P. Louineau, F. Crick, B. McNelis, R.D.W. Scott, B.E. Lord, R. Noble, D. Anderson, R. Hill, and N.M. Pearsall, *Sol. Energy Mater. Sol. Cells*, 1994, *35*, 461; (b) T. Yagiura, M. Morizane, K. Murata, K.T. Yagiura, K. Uchihashi, S. Tsuda, S. Nakano, T. Ito, S. Omoto, Y. Yamashita, H. Yamakawa, and T. Fujiwara, *Sol. Energy Mater. Sol. Cells*, 1997, *47*, 2227; (c) M. Yoshino, T. Mori, M. Mori, M. Takahashi, S.-I. Oshida, and K.Shirasawa, *Sol. Energy Mater. Sol. Cells*, 1997, *47*, 235; (d) S.R. Wenham, S. Bowden, M. Dickinson, R. Largent, N. Shaw, C.B. Honsberg, M.A. Green, and P. Smith, *Sol. Energy Mater. Sol. Cells*, 1997, *47*, 325.

212. F.R. Goodman, E.A. De Meo, and R.M. Zavadil, *Sol. Energy Mater. Sol. Cells*, 1994, *35*, 385.

213. (a) Anon., *R&D (Cahners)*, 1999, *41*(5), 9; (b) R. Dagani, *Chem. Eng. News*, 1998, July 27, 14; (c) R. Wang, K. Hashimoto, A. Fujishima, M. Chikuni, E. Kojima, A. Kitamura, M. Shimohigoshi, and T. Watanabe, *Nature*, 1997, *388*, 431.

214. N. Tsujino, T. Ishida, A. Takeoka, Y. Mukino, E. Sakoguchi, M. Ohsumi, M. Ohinishi, S. Nakano, and Y. Kuwani, *Sol. Energy Mater. Sol. Cells*, 1994, *35*, 497.

215. Anon., *Environment*, 1998, *40*(8), 21.

216. (a) M.M. Koltun, *Sol. Energy Mater. Sol. Cells*, 1994, *35*, 31; (b) T. Fujisawa and T. Tani, *Sol. Energy Mater. Sol. Cells*, 1997, *47*, 135; (c) M.D. Bazilian, F. Leenders, B.G.C. van der Ree, and D. Prasad, *Solar Energy*, 2001, *71*, 57.

217. (a) M. Brower, *Cool Energy: Renewable Solutions to Environmental Problems*, MIT Press, Cambridge, MA, 1992; *Environmental Impacts of Renewable Energy Technologeis*, Union of Concerned Scientists, Cambridge, MA, Aug. 1992; (b) K.S. Betts, *Environ. Sci. Technol.*, 2000, *34*, 306A.

218. R. Hunt and J. Hunt, *Chem. Ind. (Lond.)*, 1998, 227.

219. (a) R. Gasch and J. Twele, ed., *Wind Power Plants—Fundamentals, Design, Construction and Operation*, Stylus Publishing, Herndon, VA, 2004; (b) J.F. Manwell, J.G. McGowan, and A.L. Rogers, *Wind Power Explained—Theory, Design and Application*, 2nd ed., Wiley, New York, 2009; (c) T. Ackermann, ed., *Wind Power in Power Systems*, Wiley, New York, 2005; (d) S. Heier, *Grid Integration of Wind Energy Conversion Systems*, 2nd ed., Wiley, New York, 2006; (e) J. Deyette, *Catalyst* (Union of Concerned Scientists), 2006, *5*(2), 2.

220. J. Topping, *Chem. Eng. News*, 1997, Sept. 22, 24.

221. D. Schneider, *Am. Sci.*, 2007, *95*, 490.

222. (a) D. Tenenbaum, *Technol. Rev.*, 1995, *98*(1), 38; (b) P.M. Wright, *Chem. Ind. (Lond.)*, 1998, 208; (c) J.E. Mock, J.W. Tester, and P.M. Wright, *Annu. Rev. Energy Environ.*, 1997, *22*, 305; (d) J. Johnson, *Chem. Eng. News*, August 17, 2009, 30.

223. G. Ondrey, *Chem. Eng.*, 1999, *106*(2), 19.

224. G. Vogel, *Science*, 1997, *275*, 761.

225. A. Macdonald-Smith, *Wilmington DE News J.*, Feb. 22, 2009, C5.

226. T.B. Johansson, H. Kelly, A.K.N. Reddy, and R.H. Williams, eds, *Renewable Energy: Sources for Fuels and Electricity*, Island Press, Washington, DC, 1993, 515, 530.

227. J. Webb, *New Sci.*, 1995, July 29, 6.

228. Anon., *Science*, 1998, *280*, 1843.

229. H. Blankesteijn, *New Sci.*, 1996, Sept. 14, 20.

230. C. Flavin and N. Lenssen, *Technol. Rev.*, 1995, *98*(4), 42.

231. (a) T.B. Johansson, H. Kelly, A.K.N. Reddy, and R.H. Williams, eds, *Renewable Energy: Sources for Fuels and Electricity*, Island Press, Washington, DC, 1993, 171, 1042; (b) S.M. Schoening, J.M. Eyer, J.J. Iannucci, and S.A. Horgan, *Annu. Rev. Energy Environ.*, 1996, *21*, 347; (c) J. Fahey, *Forbes*, 2008, *182*(11), 122; (d) C. Forsberg and M. Kazimi, *Science*, 2009, *325*, 32; (e) D. Charles, Science, 2009, *324*, 172 ; (f) B.S. Lee and D.E. Gushee, *Chem. Eng. Prog.*, 2009, *105*(4), 22.

232. C. Flavin, *World Watch*, 1996, *9*(1), 15.

233. R.J. Farris, University of Massachusetts at Amherst, private communication, June, 1997.

234. (a) H. Cao, Y. Akimoto, Y. Fujiwara, Y. Tanimoto, L.-P. Zhang, and C.-H. Tung, *Bull. Chem. Soc. Jpn.*, 1995, *68*, 3411; (b) M. Maafi, C. Lion, and J.-J. Aaron, *New J. Chem.*, 1996, *20*, 559; (c) I. Nishimura, A. Kameyama, T. Sakurai, and T. Nishikubo, *Macromolecules*, 1996, *29*, 3818.

235. (a) J. Besenhard, *Handbook of Battery Materials*, Wiley-VCH, Weinheim, 1999; (b) R. Dell, *Chem. Br.*, 2000, *36*(3), 34; (c) M. Tsuchida, ed., *Macromol. Symp.*, 2000, *156*, 171, 179, 187, 195, 203, 223; (d) M.S. Whittingham, R.F. Savinell, and T. Zawodzinski, eds, *Chem. Rev.*, 2004, *104*, 4243–4886; (e) R.M. Dell and D.A.J. Rand, *Understanding Batteries*, Royal Society of Chemistry, Cambridge, 2001; (f) R. Armstrong and A. Robertson, *Chem. Br.*, 2002, *38*(2), 38; (g) T.W. Walker, *Chem. Eng. Prog*, 2008, *104*(3), S23; (h) M.R. Palacin, *Chem. Soc. Rev.*, 2009, *38*, 2565; (i) J. Chen and F. Cheng, *Acc. Chem. Res.*, 2009, *42*, 713.

236. (a) D. Sperling, *Electric Vehicles and Sustainable Transportation*, Island Press, Washington, DC, 1995; (b) S. Dunn, *World Watch*, 1997, *10*(2), 19; (c) S. Wilkinson, *Chem. Eng. News*, 1997, Oct. 13, 18; (d) J. Glanz, *New Sci.*, 1995, Apr. 15, 32; (e) G.L. Henriksen, W.H. DeLuca, and D.R. Vissers, *Chemtech*, 1994, *24*(11), 32; (f) for a magnesium battery see M. Freemantle, *Chem. Eng. News*, Oct. 16, 2000, 8.

237. N.L.C. Steele, and D.T. Allen, *Environ. Sci. Technol.*, 1998, *32*(1), 40A.
238. (a) F.-S. Cai, G.-Y. Zhang, J. Chen, X.-L.Gou, H.-K.Liu, and S.-X. Dou, *Angew. Chem. Int. Ed.*, 2004, *43*, 4212.
239. (a) J.R. Owen, *Chem. Soc. Rev.*, 1997, *26*, 259; (b) M. Jacoby, *Chem. Eng. News*, Dec. 27, 2007, 26; (c) M. Yoshi, R.J. Brodd, and A. Kozawa, *Lithion Ion Batteries: Science and Technologies*, Springer, New York, 2009.
240. (a) Anon., *Chem. Eng. News*, Jan. 12, 2009, 26; (b) N. Shirouzu, *Wall St. J.*, Jan. 11, 2008, A1.
241. (a) A. Coghlan, *New Sci.*, Apr. 1, 1995, 25; (b) J.-Y. Sanchez, F. Alloin, and D. Benraban, *Macromol. Symp.*, 1997, *114*, 85; (c) C. Roux, W. Gorecki, J.Y. Sanchez, and E. Belorizky, *Macromol. Symp.*, 1997, *114*, 211; (d) J.-F. Moulin, P. Damman, and M. Dosiere, *Macromol. Symp.*, 1997, *114*, 237; (e) F.M. Gray, *Polymer Electrolytes*, Royal Society of Chemistry, Cambridge, UK, 1997; (f) V. Chandrasebhar, *Adv. Polym. Sci.*, 1998, *135*, 139; (g) G.S. MacGlashan, Y.G. Andrew, and P.G. Bruce, *Nature*, 1999, *398*, 792.
242. J.-L. Ju, Q.-C.Gu, H.-S. Xu, and C.-Z. Yang, *J. Appl. Polym. Sci.*, 1998, *70*, 353.
243. S.-W. Hu and S.-B. Fang, *Macromol. Rapid Commun.*, 1998, *19*, 539.
244. M. Freemantle, *Chem. Eng. News*, Sept. 25, 2000, 12.
245. (a) J.D. Genders and D. Pletcher, *Chem. Ind.* (*Lond.*), 1996, 682; (b) H. Wendt, S. Rausche, and T. Borucinski, *Adv. Catal.*, 1994, *40*, 87.
246. (a) K.D. Moeller, *Tetrahedron*, 2000, *56*, 9527; (b) J. Grimshaw, *Electrochemical Reactions and Mechanisms in Organic Chemistry*, Elsevier, Amsterdam, 2001; (c) H. Lund and O. Hammerich, eds, *Organic Electrochemistry*, 4th ed., Dekker, New York, 2001; (e) K. Izutsu, *Electrochemistry in Non-Aqueous Solutions*, Wiley-VCH, Weinheim, 2002.
247. N. Takano, M. Ogata, and N. Takeno, *Chem. Lett.*, 1996, *25*, 85.
248. G. Parkinson, *Chem. Eng.*, 1996, *103*(5), 23.
249. S. Chardon-Noblat, I.M.F. DeOliveira, J.C. Moutet, and S. Tingry, *J. Mol. Catal. A: Chem*, 1995, *99*, 13.
250. M. Fremantle, *Chem. Eng. News*, June 9, 1997, 11.
251. W. An, J.K. Hong, P.N. Pintauro, K. Warner, and W. Neff, *J. Am. Oil Chem. Soc.*, 1998, *75*, 917.
252. (a) V. Ramamurthy and K.S. Shanze, *Organic Photochemistry*, Dekker, New York, 1997; (b) C.E. Wayne and R.P. Wayne, *Photochemistry*, Oxford University Press, Oxford, 1996; (c) D.C. Neckers, D.H. Volmam, and G. von Bunau, *Adv. Photochem.*, 1999, *25*, and earlier volumes; (d) A. Bhattacharya, *Prog. Polym. Sci.*, 2000, *25*, 371; (e) V. Balzani, A. Credi, and M. Venturi, *ChemSusChem.*, 2008, *1*, 26; (f) V.M. Parmon, D. Kozlov, and P. Smirniotis, *Photocatalysis—Catalysts, Kinetics and Reactors*, Wiley-VCH, Weinheim, 2009; (g) P. Klan and J. Wirz, *Photochemistry of Organic Compounds*, Wiley, Hoboken, NJ, 2008.
253. J.H. Krieger and M. Freemantle, *Chem. Eng. News*, 1997, July 7, 15.
254. G.A. Epling and Q. Wang. In: P.T. Anastas and C.T. Farris, eds, *Benign by Design*, A.C.S. Symp. 577, Washington, DC, 1994, 64.
255. T. Igarashi, K. Konishi, and T. Aida, *Chem. Lett.*, 1998, 1039.
256. B. Heller and G. Oehme, *J. Chem. Soc. Chem. Commun.*, 1995, 179.
257. J.-T. Li, J.-H.Yang, and T.-S. Li, *Green Chem.*, 2003, *5*, 433.
258. J.V. Crivello and M. Sangermann, *Polym. Preprints*, 2001, *42*(2), 783.
259. P. Zurer, *Chem. Eng. News*, Apr. 1, 1996, 5.
260. H. Fujiwara, J. Tanaka, and A. Horiuchi, *Polym. Bull.*, 1996, *36*, 723.
261. E.A.G. Gonzalez de los Santos, M.J.L. Gonzalez, and M.C. Gonzalez, *J. Appl. Polym. Sci.*, 1998, *68*, 45.
262. N.S. Nandurkar, M.D. Bhor, S.D. Samant, and B.M. Bhanage, *Ind. Eng. Chem. Res.*, 2007, *46*, 8590.

263. (a) A. Loupy, *Microwaves in Organic Synthesis*, 2nd ed., Wiley-VCH, Weinheim, 2002; (b) V. Polshettir and R.A. Varma, *Chem. Soc. Rev.*, 2008, *37*, 1546; (c) B.L. Hayes, *Microwave Synthesis—Chemistry at the Speed of Light*, CEM Publishing, Matthews, NC; (d) C.O. Kappe, *Chem. Soc. Rev.*, 2008, *37*, 1127; (e) M. Nuchter, B. Ondruschka, W. Bonrath, and A. Gum, *Green Chem.*, 2004, *6*, 128–141; (f) B.L. Hayes, *Aldrichchim. Acta*, 2004, *37*(2), 66; (g) V. Marx, *Chem. Eng. News*, Dec. 13, 2005, 14; (g) C.O. Kappe, D. Dallinger, and S. Murphee, eds, *Practical Microwave Synthesis for Organic Chemists—Strategies, Instruments and Protocols*, Wiley-VCH, Weinheim, 2008.

264. X. Querol, A. Alastuey, A.L. Soler, F. Plana, J.M. Andres, R. Juan, P. Ferrer, and C.R. Ruiz, *Environ. Sci. Technol.*, 1997, *31*, 2527.

265. C.-G. Wu and T. Bein, *Chem. Commun.*, 1996, 925.

266. (a) R.S. Varma, *Green Chem.*, 1999, *1*(1), 43; (b) R.S. Varma, K.P. Naicker, D. Kumar, R. Dahiya, and P.J. Liesen, *J. Microw. Power Electromagn. Energy*, 1999, *54*(2), 113.

267. T.D. Conesa, J.M. Campelo, J.H. Clark, R. Luque, D.J. Macquarrie, and A.A. Romero, *Green Chem.*, 2007, *9*, 1109.

268. T. Razzaq and C.O. Kappe, *ChemSusChem*, 2008, *1*(1–2), 123.

269. A. Holzwarth, J. Lou, T.A. Hatton, and P.E. Laibinis, *Ind. Eng. Chem. Res.*, 1998, *37*, 2701.

270. F.K. Mallon and W.H. Ray, *J. Appl. Polym. Sci.*, 1998, *69*, 1203.

271. R.G. Compton, B.A. Coles, and F. Marken, *Chem. Commun.*, 1998, 2595.

272. R. Roy, D. Agrawal, J. Cheng, and S. Gedevanishvili, *Nature*, 1999, *399*, 668.

273. D.E. Clark, W.H. Sutton, and D.A. Lewis, eds, *Microwaves—Theory and Application in Materials Processing IV: Microwave and RF Technology: From Science to Application*, Ceramic Trans., vol. 80, American Ceramic Society, Westerville, OH, 1997.

274. W.C. Conner and R. Laurence, Department of Chemical Engineering, University of Massachusetts at Amherst, private communication, June 1997.

5 Environmental Economics

5.1 INTRODUCTION

Some of the previous discussions have covered ways of reducing pollution by prevention. This is often cheaper than putting something at the end of the pipe or at the top of the smokestack. For example, substitution of an aqueous cleaning process for one using trichloroethylene means that no solvent has to be purchased, recycled, or disposed of. No solvent emissions result from the new process, and none can be spilled and seep into the groundwater. The used water can be treated on site or at the local wastewater treatment plant. Processes that will help society attain a sustainable future have also been discussed. According to Ed Wasserman, President of the American Chemical Society for 1999, "Green chemistry is effective, profitable and it is the right thing to do for our health and that of our planet."[1] The problem is that companies and ordinary citizens have been slow to adopt these new processes.[2] This chapter will explore what role money plays in this and what other social and political factors may be involved in the development of new policies.

Under conventional economics, market forces can destroy or profoundly affect the life support system on which we depend. This destruction is now evident in worldwide problems, such as stratospheric ozone depletion and global warming. On a smaller scale, a commercially important species can be driven to near extinction by ordinary market forces.[3] The striped bass fishery on the East Coast of the United States provided fish for the restaurant trade. As the supply of fish was reduced by overfishing, the restaurants charged more and customers were willing to pay the price. Finally, when the population was seriously depleted, the federal government initiated a moratorium on fishing for the species. After a few years without harvesting, the population began to recover.

A new hybrid discipline of environmental economics (also called ecological economics) tries to include the value of natural support systems in a consideration of profit and loss. Ecologists and economists are coming to know each other. The International Society for Ecological Economics publishes the journal *Ecological Economics*. A number of reviews on the subject are available.[4]

5.2 NATURE'S SERVICES

Nature provides a number of important services[5] free of charge. These are seldom included in economic calculations. Robert Costanza et al. estimated their total value at US$33 trillion/yr in 1997.[6] (The authors emphasize that these figures are preliminary and will require further refinement in the future.) For comparison, the world's gross national product per year was about US$18 trillion. The gross annual national product of the United States was US$6.9 trillion. The 17 services were estimated for 16 different biomes, including forest, grassland, wetlands, ocean, and so on. They included the following:

Gas regulation: composition of the atmosphere, ozone layer, and so on
Climate regulation: temperature, precipitation, and so on
Disturbance regulation: resilience to storms, droughts, and so on
Water regulation: the hydrological cycle
Water supply: storage and retention of water, aquifers, and so on
Erosion and sedimentation control
Soil formation
Nutrient cycling: nitrogen, phosphorus, and so on
Waste treatment: recovery of nutrients
Pollination
Biological pest control
Refugia: nursery grounds, overwintering grounds, and so on
Food production: fish, game, and so on
Raw materials: lumber, fuel, fodder, and so on
Genetic resources: medicines, genes for resistance to pests, and so on
Recreation: ecotourism, sport fishing, and so on
Cultural: aesthetic, artistic, educational, or scientific value

The largest value, US$17 trillion, is assigned to nutrient cycling. Because ecosystem services are often ignored or undervalued, the social costs of a project may exceed the benefits. As an example, clear-cutting a forest will produce income for the landowner, the timber company, and the workers. It will also decrease the value of nearby homes. The sediment that washes off in the process may make the streams unsuitable for spawning fish, fill up reservoirs, and kill the coral reef in the ocean where the stream enters it. When it rains, more water will run off and less will soak into the ground. Floods will occur more often. Another result may be that the stream now dries up in the summer so that the adjacent town has no water to drink or to use in irrigating crops. If these services could be replicated at all, the cost would be high. The water might be brought in by pipes from another watershed. The fish might be imported from another part of the globe. The electricity from the lost hydropower might be supplied by a new coal-burning plant.

The economic value of biodiversity has also been calculated by others.[7] Pimentel et al.[8] estimate the value of biodiversity at US$3 trillion/yr. Their list of services includes the following:

Soil formation
Biological nitrogen fixation
Crop and livestock genetics
Biological pest control
Plant pollination
Pharmaceuticals

They point out that biodiversity is necessary for the sustainability of agricultural, forest, and natural ecosystems on which humans depend. Roughly 99% of pests are controlled by natural enemies and by plant resistance. Loss of key pollinators may mean loss of a crop. Honeybees can pollinate some crops, but not all of them. There are already species for which no pollinators are left.[9] Current extinction rates are 1000–10,000 times the natural rate. The concern is that keystone species (i.e., species without which the ecosystem cannot function) may be lost. These are more likely to be soil microbes or insects rather than pandas or tigers, despite the popular appeal of the latter.

Just the extent of human alteration of the natural cycles is causing problems. Soil microbes and lightning fix about 90–140 million tons of nitrogen each year. Humans add 140 million more tons as fertilizer made by the chemical industry to increase crop yields.[10] This has accelerated the loss of biodiversity. Nitrogen is a key element in controlling species composition, diversity, and dynamics in both terrestrial and aquatic systems. Many of the original native plants function best at low levels of nitrogen. Many soils are now so saturated with nitrate that vital nutrients, such as calcium and potassium ions, are being lost to groundwater and streams. Algal blooms and toxic *Pfiesteria* outbreaks are more common. This has contributed to long-term declines in coastal fisheries. There is now, as a result of fertilizer runoff from Midwestern farms, a large area in the Gulf of Mexico with too little dissolved oxygen to support the fishery. Some of the nitrogen is released to the air as N_2O, a greenhouse gas that helps to deplete the ozone layer.

5.3 ENVIRONMENTAL ACCOUNTING

5.3.1 THE ECOLOGICAL FOOTPRINT

Each year, 6 million ha of land in the world undergoes desertification, 17 million ha are deforested, and soil erosion exceeds soil formation by 26 billion tons. The ecological footprint is an accounting tool devised by William Rees to measure the productive land area needed to supply the resource consumption and waste assimilation of a given human population.[11] Twenty percent of the world's population, largely in industrialized nations, consumes 80% of its resources. Some typical values of the productive land or water needed to support a person at a given material standard indefinitely are given in Table 5.1.

Many developed nations do not have enough productive land or water to support the current material consumption. Typical deficits are given in Table 5.2.

The water footprint of common consumer goods varies widely with typical values such as 140 L per cup of coffee, 70 L per apple, 2700 L per cotton shirt, 40 L per slice of bread, 15,500 L per kg of beef, and 8000 L per pair of leather shoes.[12]

TABLE 5.1
Typical Values of Land or Water Needed to Support a Person

Country	Land or Water Needed (ha/person)
United States	5.1
Canada	4.3
The Netherlands	3.3
India	0.4
World	1.8

TABLE 5.2

Typical Land and Water Deficits by Country

Country	Deficits (ha/person)
Japan	1.76
Korea	1.81
Austria	2.15
Belgium	2.8
United Kingdom	2.65
Denmark	2.38
France	2.22
Germany	2.66
The Netherlands	2.85
Switzerland	2.56
United States	2.29

Materials to make up the deficits are supplied by other nations.

Calculation of the ecological footprints of cities around the Baltic Sea showed that the area required for the assimilation of their waste nitrogen, phosphorus, and carbon dioxide would be 390–975 times the land area of the cities.[13] The amount of forest land needed to assimilate carbon dioxide from all the world's cities would be three times that available on earth. The World Wildlife Fund "Living Planet Index" tracks the effect of human populations on biodiversity.[14] In 2004, the biggest consumers of nonrenewable resources per capita were the United Arab Emirates, the United States, Kuwait, Australia, and Sweden.[14]

5.3.2 LIFE-CYCLE ANALYSES

Environmental life-cycle analyses follow a product from cradle to grave.[15] (Following it from cradle to cradle would be better.) They may not be easy to do and they may be expensive. They have to take into account the environmental effect at each stage: the raw materials and the way they were obtained; the manufacturing process; any transportation involved; the influence of use; and how the object will be disposed of at the end of its useful life. When no hard data from detailed studies are available, simplifying assumptions may have to be made. It is not surprising that different persons evaluating the same product can come to different conclusions. The first thing to do in looking at an analysis is to see if the group that funded it might have a vested interest in the outcome.[16] The second thing to do is to see what assumptions have been made.

Several studies have compared the environmental effects of disposable diapers with cloth diapers.[17] (The world uses 450 billion disposable diapers per year, of which 93 billion are used in India.) A 1988 analysis by the cloth diaper industry favored cloth as having a smaller environmental effect. A 1990 study made for Procter & Gamble, makers of disposable diapers, favored disposable diapers. A 1990 study by the American Paper Institute found the two kinds to be equivalent. A 1991 study by the cloth diaper industry found cloth to be superior.

*solution to the
way they are
obtained*

A disposable diaper is often made with a nonwoven polypropylene fiber face sheet and a transfer layer next to the skin backed by a layer of cellulose and a superabsorbent polymer, with a polyethylene cover on the outside. Over the years, they have become thinner. The polyethylene and polypropylene are unlikely to have been made from renewable raw materials, although it is conceivable that they might be in the distant future. The superabsorbent may or may not have been made at least partially from starch. Thus, one must assess the possibility of oil spills in drilling for and recovering the oil. What wastes were associated with this? Was any forest destroyed in the process? What pollution or wastes came out of the oil refinery? What type of forest was cut to produce the cellulose? Was old-growth forest cut? Did the logging result in sediment damaging a fishery or in other loss of biodiversity? What pollution and wastes were involved in making the superabsorbent? The acrylonitrile that may have been grafted to starch to produce the superabsorbent is a carcinogen. What was the energy needed to make the diaper and to transport it to the user? The energy probably came from fossil fuel rather than from a renewable source. What was the fuel efficiency of the equipment used in the various steps of acquiring the raw materials, making the diaper, and transporting it? After use, the diaper was probably buried in a landfill where even the cellulose part may not decompose.

The cotton diaper also involves the use of energy that probably comes from fossil fuels. Again, what was the fuel efficiency of the equipment used in the various steps in the life cycle? The cotton may have been raised with subsidized irrigation water and with pesticides and fertilizers, some of which may have seeped into the groundwater or the nearest stream. Were the workers in the cotton mill protected from dust? Because the diaper can be used many times, the main environmental effect will involve making the detergent, the washing machine, and heating the water. What wastes and pollution may have accompanied the making of the detergent? Is the energy to heat the water from a fossil fuel or a renewable source? What is the energy efficiency of the water heater? How much hot water is used? Swedish clothes washers may take only a small fraction of the water used in an automatic clothes washer in the United States. How hot is the water? Is the detergent used in warm water instead of hot water? Detergents for use in unheated water are now available commercially. Is the diaper dried on a line in the sun or in a drier powered by gas or electricity? Are two worn diapers sewn together to obtain more life from them? When the child no longer needs the diapers, will they be given to a friend who will use them until they wear out? At the end of their useful life, will the diapers be used as wiping clothes, in rag content paper, or just sent to the landfill? How do you depreciate a washing machine that is used to wash other clothes as well?

One of the studies of cloth diapers assumed a life of 92 uses (a study funded by the disposable industry), whereas the other assumed 167 uses (a study funded by the cloth diaper industry). A study funded by the disposable diaper industry assumed an energy cost for transporting the cotton to China, where the diapers were made. One would have to check to see what fraction of cloth diapers is made this way. Disposable diaper makers have pushed composting as a method of disposal. Most cities do not compost this type of waste. The polyolefins would not degrade well in this process, even though the lack of stabilizers in the polypropylene might allow it to become powder.

Cotton towels last through about 100 washes. Hotels can make them last longer by posting signs that suggest that guests not ask that they be washed every day.[18]

This comparison of diapers shows how many assumptions and judgment calls have to be made in typical life-cycle analyses. To perform a comprehensive analysis would require collecting a large amount of data that might be expensive to obtain. It also points out the possible misuse of data. A diaper that involved no use of nonrenewable materials would be preferred.

Life-cycle assessment is a technique that is still evolving. The U.S. EPA is studying the technique and trying to standardize it enough to allow meaningful ecolabels to be put on consumer items.[19] Among the effects being considered are ozone depletion, global warming, smog formation, human toxicity, noise, energy use, and nonrenewable resource depletion. Tradeoffs may sometimes have to be made (e.g., choosing the lesser influence of toxic chemical use and global warming). The process endeavors to quantify "energy and raw material requirements, atmospheric emissions, waterborne emissions, solid wastes, and other releases for the entire life cycle of a product, package, process, material, or activity." A factor that should be included is whether or not the product has been designed for easy disassembly for recycling at the end of its useful life. Many other countries have government-sanctioned ecolabeling programs, including the European Community, Canada, France, Germany, Austria, the Netherlands, Singapore, New Zealand, and Japan.[20] Although these may differ in detail, the goal is to promote products that have reduced environmental effects during their entire life cycles. They should also help consumers make informed choices in their purchases. This will help avoid goods that unintentionally promote a dysfunctional economy.

5.3.3 Cost–Benefit Analyses

Cost–benefit analyses[21] can also involve uncertainties and tradeoffs. Some of the biggest problems involve health and safety. What value should be assigned to a human life? Is it the same for a 30-year-old and a 75-year-old? Will this value be the same for a person in a developing nation as it is in a developed one? How can the cost of sickness be measured? One can add up the costs of doctor's visits, medicines, and income lost from missing work. What dollar value can be put on time lost from school or the misery of just feeling lousy? Lead in the environment can lower the IQ of children. What dollar value should be assigned to this? What lost value is assigned to a dirty lake? Does one add up the cost of buying fish instead of pulling them out of the lake, the cost of driving elsewhere to fish or swim, and the cost of cleaning up the water for drinking?

Cost–benefit analyses have been popular with some members of the U.S. Congress and polluters, who hope that new regulations can really be justified economically.[22] That the benefits are significantly larger than the cost,[23] however, has not eliminated the grumbling when the cost of the change is high. This was true with the lowering of the ozone standard from 0.12 to 0.08 ppm and reducing the size of the particulate matter to be regulated.[24] The new regulations were estimated to prevent 20,000 premature deaths per year and 250,000 sick days. The cost would be US$6.5–8.5 billion/yr versus benefits of US$120 billion. The new particulate standards could require redesign of diesel engines in trucks and buses, as well as putting new catalytic converters on wood stoves. Changes in coal-burning power plants may also be required.

Many cost–benefit analyses are of limited value. In addition to the problems of putting a value on human health, there may be simplifying assumptions and comparisons that are not valid. This has led to their value being questioned.[25]

5.3.4 GREEN ACCOUNTING

Pollution and waste represent inefficiencies of production,[26] which might be eliminated through pollution prevention.[27] For example, if the United States were as energy-efficient as Sweden or Japan, it could save US$200 billion/yr. The chemical industry in the United States spends over US$4 billion each year on pollution control, an amount equal to 2.7% of sales.[28] An additional 2.1% of sales is spent on related capital expenditures.

Antiquated accounting systems are one hindrance to the identification of opportunities for pollution prevention. The costs of waste may be hidden in overheads, research and development, marketing, product registration, and other unrelated accounts.[29] The waste and all costs associated with it should be assigned to the process and product that produces it. This may turn out to be as much as 20% of the total production cost of the product. Amoco's Yorktown oil refinery in Virginia thought its environmental costs were 3% of operations, until it checked carefully and found them to be 22%.[30] DuPont found a case where the environmental costs were 19% of the manufacturing costs. When the true cost is properly assigned, management may have a greater incentive to restudy the whole process in the search for a better one. The pollution prevention alternative may offer opportunities for increased productivity and reduced cost. It can be cheaper than any end-of-the-pipe treatment. It may eliminate the need for waste disposal, especially disposal of hazardous waste, scrubbers on exit air, need to report toxic releases, future liabilities, and contingencies, such as fines or need to remediate spills. For example, if the solvent is replaced by water, there will be no need for an air scrubber. No solvent will need to be purchased. If no toxic reagents are used, then none will be left in exit air or water. If a use can be found for the waste, if will generate revenue rather than require a disposal fee. If no waste is sent to a landfill, then none will leach out in the event of a failure of the liner and there will be no long-term liability. In addition, a clean process may help the corporate image, which may help sales of other products as well. Capital requests for pollution prevention projects will fare better in competition with other capital requests if all these factors are taken into account. Plants with green-accounting systems have roughly three times as many pollution prevention projects as those with no cost-accounting systems. The projects yielded an average savings of US$3.49 for every dollar spent.[31] A study by the Business Roundtable (a group of top executives of large corporations) concluded that pollution prevention efforts can be successful when they are incorporated into existing business procedures, rather than being treated in separate plans, and when specific approaches are not mandated.[32]

Many plants have associated costs that are not borne by the plant, but which are externalized to others. These are also referred to as societal costs. These are not included in the usual accounting system. Businesses may not be legally accountable for them. Consider, for example, the power plant that burns high-sulfur coal. Some such plants have scrubbers that take out most of the sulfur dioxide from the stack gas. Other plants were in existence when the Clean Air Act went into effect and are grandfathered so that no scrubber is required.[33] In one place along the Ohio River, an old plant puts out six times more sulfur dioxide per unit of electricity generated as a nearby newer plant. The air receiving the sulfur dioxide drifts eastward to New York and New England, where it decreases forest and crop yields and can kill some forests. Tall stacks put on some plants in the Midwest reduce the concentration of sulfur dioxide at ground level locally, but increase long-range transport. The sulfur dioxide is converted to sulfuric acid in the air. This causes corrosion of objects made of metal, limestone, marble, and sandstone. These include automobiles, which have to have an acid-resistant finish, power lines, buildings, statues, and others. In former years, sulfur oxides from metal smelters often killed all vegetation within miles of the smelter, as at INCO's smelter in Sudbury, Ontario. (After resisting cleanup for many years, INCO found that implementing a redesigned smelting process not only reduced emissions but also reduced costs, so that it made a return on its investment.[34] It sells the sulfuric acid made from the sulfur dioxide at a profit and now markets the technology to others.) In London and in Donora, Pennsylvania, high concentrations of sulfur oxides in smog killed many persons in the 1950s. There is an incentive to keep the grandfathered plants running as long as possible because they are cheaper. The cost of adding scrubbers would raise the price of electricity to customers in Ohio, but not to the point that their electricity would be more expensive than that in New England.

5.4 CORPORATIONS

5.4.1 ADDITIONAL REASONS WHY POLLUTION PREVENTION MAY NOT BE ADOPTED

Paul Tebo of DuPont has outlined his company's intention to reduce emissions and waste with the goal of zero emissions.[35] He says that proactive companies are finding that what is good for the environment is good for business and can lead to a competitive advantage. An example he gives is the company's nylon plant in Chattanooga, Tennessee, which now converts 99.8% of the raw materials to saleable products. As a result, the site's wastewater plant will be shut down and the remaining wastewater sent to the city's treatment plant, which will save US$250,000 yr^{-1}. This has been done by finding uses for byproducts that were discarded. A second example is an herbicide plant in Indonesia that will produce only one or two bags of ash each week from its solid and liquid waste. New plants are more likely to incorporate pollution prevention than old ones that are fully depreciated, especially if the products of the old ones are considered to be mature and are not growing in sales. Ecoparks, where one company's waste is another's raw material, may become more common.[36] An example is the one at Kalundborg, Denmark.[37]

     OSi Specialties, a subsidiary of Witco (now Chemtura), was nudged into action by the U.S. EPA.[38] The problem was loss of methyl and ethyl chloride in the wastewater from a process of making ethers from polyethylene oxide and polypropylene oxide. A complete study of all of the waste streams at the plant resulted in a solution that cost US$600,000 and saved US$800,000. This was done by adding a unit that converted the excess methyl chloride from the process to methanol, which could be sold. An attorney for the company says, "But if it hadn't made economic sense, we wouldn't have done it." If dimethyl carbonate could be used for the end-capping, there would be no waste salts produced and almost no wastewater.

A study done by INFORM found that the "obstacles to pollution prevention were not regulatory, technological or even economic, but primarily institutional."[39] Most companies assumed that their processes were efficient. Some pollution control departments had little knowledge of the processes that produced the wastes that they handled. Another study found that environmental managers often lack the wholehearted support of the company's business managers, so that it is hard for them to get things done.[40]

The U.S. DOE has found that seven industrial sectors—chemicals, aluminum, steel, metal casting, glass, petroleum refining, and forest products—consume 80% of the energy used in manufacturing in the United States and are responsible for 80% of all waste and pollution from manufacturing.[41] Research and development averages about 2.8% of sales for all manufacturing in the country. Aside from the chemical industry, the six other segments spend about one-third of this. These industries have relatively mature markets and technologies. Their scale requires large capital expenditures for change, and payback periods may be longer. They are characterized by conservative investment strategies and a reluctance to innovate. This has led to high pollution, high energy use,[42] and little incentive to improve. According to the DOE, the chemical industry spends more on research and development than the average manufacturer. However, most of this is said to be on product development, rather than on basic research or research to improve processes. The DOE has put up some federal money to help companies in these industries to innovate in areas where the companies are unlikely to innovate on their own, but for which the payoffs may be large for the whole country.

A company that has downsized to the point where many employees have too much work to do may have little time to do more than service existing product lines. Management may have little incentive to invest money or research in product lines that are not growing and that it considers to be mature. Too much emphasis on the bottom line next quarter may lead to a demand for unrealistically short payback times for pollution prevention projects. There have been cases where a plant manager looked good on the balance sheet by spending very little money for maintenance. That this approach may end up costing more in the long run than planned preventive maintenance may not become apparent until the plant manager has transferred to another location or has retired. For example, a small leak in the roof will do little damage if fixed promptly. If it is allowed to remain until it becomes large, the repairs may include not only new roofing but also the beams that support the roof, the ceiling and floors in the room below, and any equipment in the rooms that would be damaged by water. Management may also be wary of a pollution prevention project that it feels falls under someone else's patent. It may feel that any royalty charged might be enough to negate any cost savings that might result from the project. It may also look at end-of-the-pipe treatments as simpler and easier than reexamining their whole process of finding ways to improving it. There may also be the fear that the new process will not pay for itself in savings, with the result that competitors who do not clean up will have an advantage in the marketplace. If management feels that a better environmental technology is just around the corner, it may fear locking into a process that might be obsolete in a few years.

5.4.2 Other Aspects of Corporate Finance

In a fire or explosion, there is the cost of rebuilding the plant, the possible loss of business to competitors during the period when your plant cannot supply, hospital and rehabilitation bills for the injured, government fines for safety violations, awards by juries for damage to the community, and loss of a good company image that may have helped to sell your products. This should provide a powerful incentive for substituting greener processes that use less toxic chemicals. In addition to the reasons for accidents in Chapter 1, there may also be the feeling that "It can't happen here. We've been doing it this way for years and have never had a problem."

Corporations are often said to place their interests over those of the community, the state, and the environment. It is to their financial advantage to externalize as many costs as possible rather than internalize them. After the logging company removes the timber in some countries, replanting may be left to the government. A generation or two ago in the United States, beer and soft drinks were sold in refillable bottles with deposits on them. Since then, the bottlers have turned to single-use throwaway containers. The washing lines have been dismantled and their crews laid off or shifted to other jobs. The consuming public pays more now for the packaging and for its disposal. The local government usually provides the landfill or incinerator to deal with the packaging after it is used. Firms may tend to overexploit costless inputs such as air and water, in what Hardin called the "tragedy of the commons."

Little work is done on drugs for major tropical diseases, because the pharmaceutical industry fears that the persons who contract the illnesses will be too poor to purchase the drugs.[43] However, some drug companies have sold drugs of questionable benefit in developing countries.

Many firms that make money by supplying fossil fuels are reluctant to diversify into energy-supply companies. They could also make money by selling and servicing equipment and techniques of energy conservation or by selling energy from renewable sources. However, the techniques for obtaining energy from renewable sources may be unfamiliar to them. Electric power plants could sell more energy if they used their waste heat for district heating and cooling instead of putting it into a river. Some public utilities commissions in New England have worked out systems for rewarding utilities for selling energy conservation to their customers so that less electricity needs to be generated.

A firm in the business of selling agricultural chemicals might diversify into integrated pest management. This would mean fewer employees in chemical plants, but many more employees as ecologists, entymologists, soil scientists, and such to monitor insect populations, crop damage, fertilizer needs, or others. Such firms could sell consulting services. They might also obtain patents on insect traps, pheromones, or insect repellents, on which they could obtain royalties on the use by others.

These examples show a switch from pure manufacturing to broader companies that provide a variety of services as well as products. Unfortunately, large firms are less innovative than small ones and are often reluctant to make the kinds of changes suggested here. The changes would involve the retraining of workers no longer needed in production, but the new company structure would require more total employees. As far as consumer products are concerned, it often takes more labor to repair one than to produce a new one by mass production methods.[44] This overlooks the cost of disposal.

American companies are becoming greener as more of them embrace the "triple bottom line" of profit, environment, and social responsibility.[45] Companies with strict standards are finding that their stock market valuation has improved and that it is easier to get loans and insurance.[46] Socially responsible investing is growing.[47] Companies are being rated on their "triple bottom line" by Innovest Strategic Value Advisors, Sustainable Asset Management, and the Equator Principles Financial Institutions.[48]

5.5   ENVIRONMENTAL ECONOMICS OF INDIVIDUALS

5.5.1   MAKING CHOICES

Individuals try to minimize their expenses, but they do not always buy or do what is cheapest, at least not what is the cheapest per unit of performance, or cheapest over the long term. They sometimes justify this in terms of increased "convenience." Some of this is a matter of what dollar value they place on their time that is not spent at work earning money. The leaf blower is much more expensive and uses more materials and fossil fuel than a rake that is used by hand. The rake can also provide needed exercise. Yet, they seem to be increasing in popularity. The use of single-use throwaway beverage containers, which are more expensive than refillable ones on a per trip basis, was mentioned earlier. A similar situation exists in fast-food restaurants. Throwaway dishes and eating utensils cost more per use than washable, reusable ones. They are popular with the restaurant management because no equipment, labor, or space is required for washing them. Consumers often buy less efficient items, such as fuel-inefficient vans, justifying their purchases by "convenience" and "I can afford it." They may also buy for style and fashion, rather than longer life and better wear.

There are often hidden costs not seen by the consumer when he or she makes his or her choice. Stores in America create the illusion of abundance and perpetuate the myth of plenty. Few consumers see the natural resource depletion and unsustainable nature of this. The environmental abuse that created the object may be well out of sight, thousands of miles away. Buying a teak bench or tropical hardwood paneling may involve destruction of tropical forests by unsustainable harvesting. Wild flowers are popular with home gardeners. Buying them may contribute to the decimation of natural populations if the plants are collected from the wild. The situation is not helped by nurserymen who collect from the wild and then grow the plants in their nurseries for a year and label them as "nursery grown." Ecolabels will help consumers make informed choices only if the labels are honest. This may require certification by an unimpeachable agency, possibly by government.

The purchase of food may involve similar hidden costs. The food may have been raised on land that formerly was a diverse forest or prairie. It may have involved the heavy use of pesticides and soil loss. Because the average food item consumed in the United States has been shipped over a thousand miles, the energy cost in distributing it is high. Buying local produce would reduce this energy cost. Without informative labels, how can the shopper tell the difference? People may inadvertently contribute to the loss of local farm products by buying a new house on what may have been some of the best farmland in the state. This will eventually raise food prices, because the yields from marginal land will be lower or food will have to come from farther away. People who buy waterfront lots, complete with a boat slip at the back of the house, may not think about the productive wetland that was replaced, but may be disappointed at the decline in the fish populations that they came to catch. The abuser may be far away from the problem that he or she helped to create. A farmer in Lancaster County, Pennsylvania, who uses a lot of fertilizer in the hope of maximizing his crop yields may not realize that he is contributing to the decline of life in the Chesapeake Bay downstream (i.e., unless he goes to the Bay to fish or sail in his off time).

5.5.2 ECOTOURISM

Ecotourism is becoming big business.[49] Two journals are devoted to this subject.[50] If done properly, it can be an important way of preserving biodiversity and bringing income to the area. It is the top source of income for Costa Rica, which is doing it in a sustainable fashion.[51] It is the first developing country to halt and then reverse its deforestation. The country uses a tax on fossil fuels to pay rural landowners US$50 ha/yr to protect their forests and US$400 ha/yr to farm ecologically. It has also achieved zero population growth at a population of 4 million.

However, there are limits. The actions of tourists must not be so much that they degrade the area, or tourists will stop coming.[52] (Certification by a government or another reputable agency helps to minimize these problems.[53]) In Sri Lanka, the coast was degraded by the lack of proper disposal means for garbage and wastewater.[54] Trekking in the Himalayan Mountains has resulted in deforestation and litter accumulation along trails.[55] Tourism in Bali has resulted in the fragmentation and degradation of coastal ecosystems of coral reefs, seagrass, and mangroves.[56] The message in all of these cases is that proper service facilities must be provided for the tourists who come. The amount and extent of resource use should be planned in advance by the community and the businesses there. Recreational diving in eastern Australia involved the breakage of 0.6–1.9 corals per 30-min dive.[57] The bulk of the damage was done by a small number of inexperienced divers. Predive briefing helped reduce the problem. Because the most damaging activity of all was the anchoring of the boats that brought the divers, putting in established mornings should help.

5.6   GOVERNMENT ACTIONS AFFECTING ENVIRONMENTAL ECONOMICS

5.6.1   ROLE OF GOVERNMENT

The government of Bhutan has suggested replacing the usual gross domestic product (GDP) per capita per year with a gross national happiness (GNH).[58] The rating is based on (1) promotion of equitable and sustainable socioeconomic development, (2) preservation and promotion of cultural values, and (3) conservation of natural resources and establishment of good governance. Other rating systems that proposed a replacement for GDP per capita per year are Genuine Progress Indicator, Happy Planet Index, and Human Development Index.

The interplay of the marketplace can allow results to happen that are not in the long-term interest of society. For example, it can drive a species to extinction. It can so degrade the "commons" of air and water that they can no longer provide the free natural services that they once did. Government can provide a series of incentives to encourage individuals and businesses to manage these resources in a sustainable way.[59] It can set up accreditation systems to certify that a product has been produced in an environmentally sound fashion (e.g., a green label indicating that the tropical wood has been harvested in a sustainable manner).[60] Germany has used a government-regulated "Blue Angel" logo on environmentally sound products, such as mercury-free thermometers, for the past 20 years.[61] Voluntary labeling of wood grown on a sustainable basis and crops produced by organic farming methods is being tried.[62] It is endorsed by environmental organizations. Such systems need independent verification of compliance. Such certification may offer some competitive advantage.

Most of the current regulations in the United States are based on considerations of health and safety.[63] In some cases, it has been necessary to use a precautionary principle and act when the problem is definite, but before the final scientific evidence is in, as in the Montreal Protocol for substances that deplete the ozone layer.[64] As problems of ozone depletion, global warming, and loss of biodiversity have reached global dimensions, international treaties among many nations have become necessary to address these problems.[65] International institutions, such as the World Bank, as well as governments are working harder to prevent further damage to our life support system, the earth.[66]

The State Environment Protection Inspectorate of Poland is trying to improve Poland's poor environmental record by a "name and shame" policy of publishing a list of the country's polluters each year.[67] Since 1990, the list has dropped from 80 to 69.

Government can establish laws to restrict abuses. If something is terribly bad, it can be banned completely, as occurred in the United States in the banning of seed treatments involving organomercury compounds. Mercury is involved in enough fish advisories that setting a limit of zero on wastewater may be desirable. This would cause a shift of any remaining chlorine plants using mercury electrodes to the well-proved method using Nafion-containing cells. (Taiwan had seven plants still using the mercury process in 1999. One of them made the news in 1999 when Formosa Plastics shipped 3000 metric tons of mercury-containing waste to Cambodia, where it was rejected.[68] Attempts to have it treated in California failed, so that the waste went back to Taiwan, where the people do not want it either. Taiwan has been slow to adopt adequate controls on environmental pollution.) Putting a zero limit on toxic heavy-metal ions in wastewater might allow sewage sludge to change from something that is hard to get rid of to a profitable soil amendment. A law proposed in Vermont would require labeling of consumer products containing mercury to help in recycling them.[69] Manufacturers of fluorescent lamps oppose this law.

5.6.2 REGULATIONS

Regulations[70] have proved to be quite valuable in curbing some of the worst pollution of air and water from point sources. They have been less effective in handling pollution from nonpoint sources. Just the threat of regulation can cause companies to study ways of reducing their pollution. Many efficient, profitable ways of reducing pollution tend to surface after regulations come into effect.[71] Regulations can reduce a firm's resistance to change. They can also provide a level playing field where a company that cleans up at some cost is not penalized in the marketplace by losing market share to one that has not. Penalties for noncompliance must be high enough to hurt the company's bottom line, so that they are not just considered as part of doing business. (Some businesses that would prefer to avoid the costs of complying with regulations quote the costs without mentioning the benefits and call for "regulatory reform.")[72]

Industry in the United States has argued for greater flexibility in correcting the problems than is allowed in some of the command-and-control laws. It says that this will allow it to seek the least-cost solutions to problems. This has started to happen. The U.S. OSHA has now invited each of the companies with the worst safety records to implement an improved safety and health program, with the guarantee that there will be significant worker involvement.[73] A pilot study of this method in Maine lowered the injury rate by 30%. However, on a national scale, the program has been halted by a court ruling in a suit brought by the National Association of Manufacturers.[74] The court ruled that OSHA did not follow the proper procedures in implementing the plan.

"Cap and trade" systems allow the polluter more flexibility in finding the least-cost way to comply.[75] The government puts a cap on the amount of a pollutant that can be emitted (at a lower level than found at the time). For sulfur dioxide emissions in the United States, this was set at half of the former level. A company that cleaned up to more than this level was given credits to sell to one that could not meet the 50% level. This led to the desired reduction of emissions 30% ahead of schedule at one-tenth the cost of some industry predictions.[76] This involved the use of low-sulfur coal instead of adding scrubbers, and should offer an additional incentive to improve methods for reducing the sulfur levels in mined coal. This method has been proposed for tackling the problem of global warming.[77] An alternative is a carbon tax on emissions. A "cap and trade" system is being used in northern New Jersey to limit the amount of mercury entering the Pasaic River.[78] A proposal has been made to lump all the pollutants from one site in a single trading credit instead of having separate credits for each one.[79] These and other market-based environmental tools have been endorsed widely.[80] For the protection of biodiversity, these include disincentives such as access fees, user fees, and noncompliance fees, as well as cost sharing of reserves, individual fishing quotas, trading rights, and such.

5.6.3 Jobs and Regulations

Hueting[81] mentions three myths in the environmental debate that must be reversed for progress to be made: (1) the environment conflicts with employment; (2) production must grow to create financing; and (3) it is too expensive for society to save the environment. There is no evidence that environmental regulations in the United States have harmed the economy.[82] Deposit–refund legislation in 10 states in the United States has created jobs, for example, 4684 in Michigan, 3800 in New York, 1800 in Massachusetts, and 350 in Vermont.[83] States with stronger environmental programs outperformed states with weaker programs. During 1965–1990, levels of carbon monoxide, sulfur dioxide, and lead were reduced in southern California. During this same period, employment grew 50% above the national average, wages were above the national average, and so was the local economy. Nations with the most stringent environmental regulations show the best economic performance. Plants with poor environmental records are no more profitable than cleaner plants in the same industry.[84] There is no evidence that superior environmental performance puts a company at a disadvantage in the marketplace. Only about 0.1% of layoffs are due to environmental regulation. The relocation of jobs offshore is mainly due to lower labor costs, not to environmental regulation. A shift to a sustainable economy is expected to create more jobs in energy efficiency, recycling, and public transportation than will be lost in the oil and coal industries, car manufacturing, and waste disposal. Other jobs will appear in wind energy, photovoltaic cells, bike path construction, and so on.[85]

Spending on environmental technologies creates jobs in part because the business is labor-intensive and often uses capital goods that are produced domestically.[86] There are more than 100,000 environmental technology firms in the United States.[87] They employ 1.3 million people and make US$180 billion each year. Their exports are US$16 billion/yr and provide a US$9.3 billion trade surplus. World markets for the environmental technology industries were US$295 billion in 1992.[88] The industry is driven by regulation. Its growth may slow down, at least in the industrialized nations, as major sources of pollution are brought under control. The next step will be to devise new processes that meet the environmental and productivity needs of the industry.[89] As an example, this includes jobs created when a new business is set up to make products from recycled materials.[90] It can also include a new machine that does a better job (e.g., the Maytag front-loading washing machine that handles larger loads in a gentler fashion than the usual top-loading machine, while saving water and electricity).[91] Further work is needed to reduce the price, which at present is twice that of the top loader.

5.6.4 Subsidies

The market is distorted by the many subsidies provided by governments. Subsidies more than US$0.95–1.95 trillion/yr are given to environmentally destructive activities.[92] Many of these subsidies shore up declining extractive industries. Coal mining in

Germany is an example. In the United States, these include below-cost timber sales in the national forests, cheap mining claims, irrigation water provided by dams built at public expense, grazing fees for public lands that are too low for sustainable use, price supports for crops raised in monoculture, highway construction, and policing costs not assigned to the automobile.[93] Those who use natural capital should be expected to pay the full cost of it,[94] as suggested in a report on the United States written by the U.N. Organization for Economic Cooperation and Development. If they did, they might use less of it, and recycled materials could compete better in the markets. The National Recycling Coalition in the United States urges the elimination of subsidies so that recycled materials can compete better with virgin materials.[95] The price of petroleum in the United States is artificially low, which encourages its wasteful use. Worldwide subsidies to fossil fuels amount to US$58–300 billion/yr.[96] The subsidy to fossil fuels in the United States is US$25 billion/yr.[97] The subsidy for water for western agriculture in the United States is US$4.4 billion/yr.[98] The U.S. Department of Agriculture also pays farmers to keep 28 million acres of marginal crop land out of production in a Conservation Reserve Program.[99] Agriculture in developed nations receives subsidies of US$311 billion/yr.[100] Subsidy-driven overcapitalization of the world's fishing fleets has resulted in drastic declines in many fish populations.[101] National subsidies to fishing in the North Atlantic Ocean have been US$2.5 billion/yr.

5.6.5 TAXES

Green taxes may be the least costly way to a sustainable future.[102] Taxes on waste, on pollution, and on environmentally damaging products, such as pesticides, fertilizers, motor vehicle fuels, as well as unsustainable natural resource depletion, have been proposed.[103] Such ecotaxes are being tested in Europe.[104] Many of these taxes are revenue neutral, the money being put back into the same industries in the form of reductions in other taxes. The amount put back can favor those industries with the most efficient processes that are producing the least pollution. These include taxes on carbon dioxide, sulfur dioxide, and nitrogen oxide emissions; lead in gasoline, fertilizer and batteries in Sweden; toxic waste and water pollution in Germany; water pollution and household waste in the Netherlands; carbon dioxide in Norway; and water pollution in France. Other taxes are also becoming more common. Denmark taxes cars and trucks by weight, which is often related to fuel efficiency. It also has a tax on disposable beverage containers, plates, cups, and cutlery. There are also taxes on paper and plastic carrier bags, pesticides in containers smaller than 1 kg, nickel–cadmium batteries, and incandescent light bulbs. Belgium has a tax on disposable razors that are not recycled, as well as one on disposable beer bottles. In Australia, products made entirely from recycled paper are exempt from sales tax. All of these are designed to shift taxes away from labor and capital and on to energy and the environment. The taxes appear to be working.[105] Taxes on heavy-metal emissions in the Netherlands have cut emissions of mercury by 97%. Iowa has reduced fertilizer use by a tax on it.[106] Similar taxes have been proposed to cut fertilizer use in Europe.[107] Denmark's waste tax reduced waste by 26% from 1987 to 1997, with recycling going up at the same time.[108] England is studying the use of ecotaxes to improve the environment.[109] There is a movement to make the ecotaxes uniform over the European Union.[110] China has proposed taxes on wastewater, noise, solid waste, and low-level radioactive waste.[111]

PROBLEMS

5.1   To Veto or Not

Your job as the governor of a state with a budgetary crisis is to protect the health and welfare of the state citizens. You believe that this can best be accomplished by stimulating the economy. The legislature has been fiscally irresponsible in sending you bills with hefty price tags. The latest one to arrive on your desk calls for measuring chemicals in the bodies of the citizens.

The people that you see around you don't appear to be terribly sick. The bill may be another example of chemophobia and fear-mongering. The American Chemistry Council assures you that its member companies are doing their best to make safe chemicals, and to keep chemicals out of people. You feel that the best way to proceed is to provide more jobs so that more people can afford to buy health insurance. Therefore, the money needed for the extensive monitoring of chemicals might better be spent for a subsidy for a new greyhound track. This would provide new jobs and bring in money—not only for your state, but also from the surrounding states. The spillover effect would stimulate hotels and restaurants nearby. Your popularity rating has been falling, since with the uncooperative legislature, you have not been able to solve the fiscal crisis any better than your predecessor. You are up for reelection next year and will need the unreserved support of the chemical industry. Some environmentalists may be alienated, but they seem to be irate most of the time. What should you do?

5.2   An Issue at Sea

You are the captain of a tanker carrying 77,000 t of bunker oil to Europe. A leak develops in the ship while it is off the coasts of three European countries. It is impossible to fix the leak at sea, and each country refuses to let the ship land to fix the leak. The oil cannot be transferred to another vessel fast enough to empty the ship, resulting in the spilling of much more oil than would have been lost if the ship had been allowed to come to port. What can you do? How can such problems be avoided in the future? Most of the oil (76 million gallons/year) in American coastal waters does not originate in tanker accidents. Where does it come from? What can be done to reduce this amount?

5.3   The Precious Plant

Alnus maritima (also called Seaside Alder) is a plant in the Betulaceae family found only in Oklahoma, Georgia, Maryland, and the marshes of coastal Delaware. Although it is not an endangered species, its distribution is quite limited. Recently, Alnusin III (0.01% of the dry weight) has been isolated from the outer surfaces of the plant roots. When administered to patients with AIDS, the disease abates and disappears in a few months. However, it takes two trees for a single treatment of a patient. If you harvest *Alnus maritima* for this purpose, the species would soon become extinct and the cure for AIDS would be lost. What policy should you adopt to preserve both the plant and the people?

5.4 Compound X

Compound X is known to cause cancer, asthma, diabetes, liver disease, and death of nerves when given to animals in large doses. It has been used widely and has become ubiquitous in the environment. Bioaccumulation occurs as it goes up the food chain. Your job as a regulator for the government is to set permissible limits for the material in food, air, and water so that no one gets affected by Compound X. How would you go about doing this? What tests need to be run?

5.5 A Valuable Community Asset

Your petroleum refinery produces much of the gasoline and fuel oil for most of the surrounding area. It is quite profitable and is currently running at capacity. However, the neighbors are worried about seven fires over the last year. As the plant manager, it is your job to calm their fears and assure them that the plant will try to do better. If you don't succeed, it could result in no bonus for the year or even loss of your job. One approach would be to highlight the jobs that the refinery provides, and how much money the plant puts into the local economy, as well as the important products that satisfy local needs. Then, you can assure the neighbors that the refinery will try to improve. However, there are some questions on your mind. Can the work be done without stopping production? Would you have to bring engineers from outside the company to perform leak detection and find the causes of the fires? Time is of the essence to lessen the protests of the neighbors. What should you do?

5.6 Fast Food Litter

Deposit-refund legislation in eight states has reduced the cans and bottles along roadsides. A national bottle drive is proposed each year in the United States Congress, but it never passes. It could save a tremendous amount of raw material. Fast food restaurant cups and wrappings are a major component of roadside litter. Devise a way to prevent this, as the "Adopt A Highway" and "Keep America Beautiful" programs are not taking care of it.

5.7 Too Many Cars

World petroleum production has peaked and global warming is evident, together with more intense storms. You are the vice president in charge of facilities and services at a university with about 20,000 students, in a city with around the same number of residents. Your university has 9000 parking spaces, which cannot satisfy current demand, even though a new parking garage was erected last year. Each house near the campus seems to have three cars in the driveway, and sometimes it is necessary to move three of them to get a fourth one out.

There ought to be a better, cheaper way to keep from having to build more garages, as well as to alleviate traffic congestion. Emory University offers financial incentives for carpools, van pools, and rapid transit riders. Carpoolers can receive a parking permit for US$100, half of the normal cost. Van poolers receive free parking and an additional subsidy. Emory also offers a discounted transit fare card to over 1000 employees. Car sharing systems (e.g., ZipCar) have also worked well at universities—they are significantly cheaper than the annual cost of owning and operating a car. Which of following policies could be implemented at your university?

1. Give the faculty and staff a 10% raise if they set a good example.
2. Give the students a 10% discount on tuition if they do not bring a car on or near the campus.
3. Add another parking garage.
4. Build a high-rise apartment building on campus so that students would be close to their classes.
5. Deny scholarships and university employment to any student with a car.
6. Double the number of free university and city buses on or near the campus. Provide discounted fare cards on regional buses and trains.
7. Extend the university to a 15 mile radius for commuting students and employees.
8. Crisscross the campus and city with new bicycle paths, preferably ones that are separate from the streets. Put up more bicycle racks on the campus and in the heart of the city.
9. Require a special, expensive permit for any car on major city streets with only one person in it.
10. Implement more local taxis with large discounts for students.
11. Send a letter separately to parents and students when a student applies for financial aid and when a student applies for a job on the campus. Testimonials from students who get around fine without a car might be used, or featuring faculty bicyclists in the student newspaper may help.

5.8 A Problem with Paper

You are a franchisee for Fatto's Burger Palace in a small college town ("where friends meet to eat"). The specialties of the house are the "Giant Burger Supreme" and the "Hound Doggie," both of which contain 26% or less fat. Business is good, especially when the college dietician decides to put soy stew on the menu in the dining halls.

Recently, agitation by environmental extremists has forced the restaurant to abandon the convenient clam shell boxes of polystyrene in favor of paper boxes. The latter are a little more expensive, but take up less space. They are somewhat difficult and time-consuming to fold. Customers are now complaining about the oil stains on the inside of the burger box. Water tends to seep through and make the package limp in some cases. This has required putting more napkins in each order. In addition, the blazing red Fatto's symbol on the top of the boxes now has fuzzy edges, whereas it was sharp and crisp on the clam shell boxes.

The environmental fanatics are still not satisfied. Yesterday, they showed up at the height of the lunch hour and dumped a pile of Fatto's boxes that they had picked up in parking lots and along roadsides on the door step of the restaurant. They did this right in front of a newspaper photographer. What can you and the supplier do to improve the situation?

5.9   Should TOSCA Be Replaced or Revised?

TOSCA (the Toxic Substances Control Act) was enacted in the aftermath of the 1984 Bhopal disaster in India (see Problem 1.1). It requires companies that produce or use any of a list of toxic compounds to report the amounts each year. It has caused some companies to reduce the amounts used. Companies do not like seeing their names in the headlines as the worst polluters in the state. Despite this success, people are now pushing for revision of the Act. Why?

5.10   A Downward Spiral

Assume that you are the United States Secretary of State or an equivalent person in another industrialized nation. Countries are using metals and nonmetals in various manufacturing processes with reckless abandon, as though they will last forever. However, some of them will last for only 10–15 more years. This will mean that nations that are now prosperous could eventually lose some of their wealth. Select a metal and a nonmetal that are essential for life as we know it, and suggest what can be done to avoid this problem.

REFERENCES

1. R. Dagani, *Chem. Eng. News*, July 5, 1999, 38.
2. (a) L. Ember, *Chem. Eng. News*, Mar. 20, 1995, 6; (b) B. Dalal-Clayton, *Getting to Grips with Green Plans—National Level Experience in Industrial Countries*, Earthscan, London, 1996.
3. (a) C.H. Freese, *Wild Species as Commodities—Managing Markets, and Ecosystems for Sustainability*, Island Press, Washington, DC, 1998; (b) T.M. Swansoon, ed., *The Economics and Ecology of Biodiversity Decline: The Forces Driving Global Change*, Cambridge University Press, New York, 1998; (c) S. Iudicello, M. Weber, and R. Wieland, *Fish, Markets and Fishermen*, Island Press, Washington, DC, 1999; (d) E. Bulte, R. Damania, L. Gillson, and K. Lindsay, *Science*, 2004, *306*, 420.
4. (a) C.D. Kolstad, *Environmental Economics*, Oxford University Press, Oxford, 1999; (b) S.L. Pimm, *The World According to Pimm: A Scientist Audits the Earth*, McGraw-Hill, New York, 2001; (c) R.N. Stavins, *Economics of the Environment: Selected Readings*, 4th ed., W.W. Norton, New York, 2000; (d) J.J. Rao, *Chem. Eng. Prog.*, 2001, *97*(11), 38; (e) E.A. Davidson, *You Can't Eat GNP—Economics As If Ecology Mattered*, Perseus Publishing, Cambridge, MA, 2000; (f) A. Gilpin, *Environmental Economics: A Critical Overview*, Wiley, New York, 2000; (g) D. Pearce and E.B. Barbier, *Blueprint for a Sustainable Economy*, Earthscan, London, 2000; (h) H.E. Daly and J. Farley, *Ecological Economics—Principles and Applications*, Island Press, Washington, DC, 2003; (i) N.O. Keohane and S.M. Olmstead, *Markets and the Environment*, Island Press, Washington, DC, 2007; (j) C.S. Russell, *Applying Economics to the Environment*, Oxford University Press, Oxford, 2001; (k) L.R. Brown, *The Earth Policy Reader*, Earth Policy Institute, Washington, DC, 2002 (www.earth-policy.org/Books/index.htm); (l) T.L. Cherry, S. Kroll, and J.F. Shogren, *Environmental Economics—Experimental Methods*, Taylor and Francis, Boca Raton, FL, 2007.
5. (a) G.C. Daily, ed., *Nature's Services—Societal Dependence on Natural Ecosystems*, Island Press, Washington, DC, 1997; (b) F. Hinterberger, E.F. Granek, S. Polasky, S. Aswani, L.A. Cramer, D.M. Stoms et al. *Ecol. Econ.*, 1997, *23*(1), 1; (c) G.C. Daily and K. Ellison, *The New Economy of Nature, The Quest to Make Conservation Profitable*,

Island Press, Washington, DC, 2002; (d) G. Heal, *Nature and the Marketplace, Capturing the Value of Ecosystem Services*, Island Press, Washington, DC, 2000; (e) R. Costanza, H. Daly, C. Folke, P. Hawken, C.S. Holling, A.J. McMichael, D. Pimentel, and D. Rapport, *BioScience*, 2000, *50*(2), 149; (f) K. Ellison, *Conservation in Practice*, 2005, *6*(3), 38; (g) J. Withgott, Science, 2004, *305*, 1100; (h) R.S. Farrow, C.B. Goldberg, and M.J. Small, eds, *Environ. Sci. Technol.*, 2000, *34*, 1381–1461; (i) C. Kremen, J.O. Niles, M.G. Dalton, G.C. Daily, P.R. Ehrlich, J.P. Fay, D. Grewal, and R.P. Guillery, *Science*, 2000, *288*, 1828; (j) E.B. Barbier, E.W. Koch, B.R. Silliman, S.D. Hacker, E. Wolanski, J. Primavera et al., *Science*, 2008, *319*, 321; (k) M. Palmer, E. Bernhardt, E. Chornesky, S. Collins, A. Dobson, C. Duke, B. Gold et al., *Science*, 2004, *304*, 1251.

6. (a) R. Costanza, R. d'Agre, R. deGroot, S. Farber, M. Grasso, B. Hannon, K. Limburg, S. Naeem, R.V.O. Neill, J. Paruelo, R.G. Raskin, P. Sutton, and M. van den Belt, *Nature*, 1997, *387*, 253; (b) J.N. Abramovitz, *World Watch*, 1997, *10*(5), 9; (c) W. Roush, *Science*, 1997, *276*, 1029; (d) S.L. Pimm, *Nature*, 1997, *387*, 231; (e) M. Rouhi, *Chem. Eng. News*, June 30, 1997, 38; (f) D. Pearce, *Environment*, 1998, *40*(2), 23.

7. (a) D. Pearce and D. Moran, *The Economic Value of Biodiversity*, Earthscan, London, 1994; (b) E.B. Barbier, J.C. Burgess, and C. Folke, *Paradise Lost? The Ecological Economics of Biodiversity*, Earthscan, London, 1994; (c) T.M. Swanson, ed., *The Economics and Ecology of Biodiversity Decline*, Cambridge University Press, Cambridge, 1995; (d) R. Baker, *Saving All the Parts—Reconciling Economics and the Endangered Species Act*, Island Press, Washington, DC, 1993; (e) D.M. Roodman, *The Natural Wealth of Nations: Harnessing the Market for the Environment*, W.W. Norton, New York, 1998; (f) A. Duraiappah, Policy Brief on "Putting the Right Price on Nature: Environmental Economics," U.N. Environment Program, Geneva, Switzerland 2006, www.scidev.net/biodiversity.

8. (a) D. Pimentel, C. Wilson, C. McCulllum, R. Huang, P. Dwen, J. Flack, Q. Tran, T. Saltman, and B. Cliff, *Bioscience*, 1997, *47*, 747; (b) D. Pimentel, L. Lach, R. Zuniga, and D. Morrison, *BioScience*, 2000, *50*(1), 53 [costs of non-native species in the U.S.].

9. S.L. Buchmann and G.P. Nabhan, *The Forgotten Pollinators*, Island Press, Washington, DC, 1996.

10. P.M. Vitousek, J.D. Aber, R.W. Howarth, G.E. Likens, P.A. Matson, D.W. Schinder, W.H. Schlesinger, and D.G. Tilman, *Ecol. Appl.*, 1997, *7*, 737–750.

11. (a) W. Rees and M. Wackernagel, *Our Ecological Footprint: Reducing Human Impact on the Earth*, New Society, Philadelphia, 1996; (b) M. Weckernagel and W.E. Rees, *Ecol. Econ.*, 1997, *20*, 3.

12. M. Voith, *Chem. Eng. News*, Oct. 6, 2008, 12.

13. C. Folke, A. Jansson, J. Larsson, and R. Costanza, *Ambio*, 1997, *26*(3), 167.

14. www.panda.org.

15. (a) B.W. Vigon, D.A. Tolle, B.W. Cornaby, H.C. Latham, C.L. Harrison, T.L. Boguski, R.G. Hunt, and J.D. Sellers, *U.S. EPA Risk Reduction Engineering Laboratory. Life Cycle Assessment—Inventory Guidelines and Principles*, Lewis, Boca Raton, FL 1994; (b) G.A. Keoleian, D. Menerey, B.W. Vigon, D. A. Tolle, B.W. Cornaby, H.C. Latham, C.L. Harrison, T. Boguski, R.G. Hunt, and J.D. Sellers, *Product Life Cycle Assessment to Reduce Health Risks and Environmental Impact*, Noyes, Park Ridge, NJ, 1994; (c) S. van der Ryn and S. Cowan, *Ecological Design*, Island Press, Washington, DC, 1995, 90–98; (d) J.S. Hirschhorn, *Chemtech*, 1995, *25*(4), 6; (e) J. Nash and M.D. Stoughton, *Environ. Sci. Technol.*, 1994, *28*, 236A; (f) T.E. Graedel, *Streamlined Life-Cycle Assessment*, Prentice-Hall, Paramus, NJ, 1998; (g) D. Ciambrone, *Environmental Life Cycle Analysis*, Lewis, Boca Raton, FL, 1997; (h) T. Krawczyk, *Int. News Fats Oils Relat. Mater.*, 1997, *8*, 266; (i) J. Kaiser, *Science*, 1999, *285*, 685; (j) C. Hendrickson, A. Horvath, S. Joshi, and L. Lave, *Environ. Sci. Technol.*, 1998, *32*, 184A; (k) S.T. Chubbs and B.A. Steiner, *Environ.*

Prog., 1998, *17*(2), 92; (l) J.J. Marano and S. Rogers, *Environ. Prog.*, 1999, *18*, 267; (m) M.A. Curran, ed., *Environ. Prog.*, 2000, *19*, 61–145; (n) M.Z. Hauschild, *Environ. Technol.*, 2005, *39*, 81A; (o) A. Tukker, *Environ. Sci. Technol.*, 2002, *36*, 71A.

16. D. Hanson, *Chem. Eng. News*, Dec. 21, 1998, 26.

17. (a) C. Crossen, *Tainted Truth: The Manipulation of Fact in America*, Simon and Schuster, New York, 1994, 140–143; (b) Anon., *World Watch*, 2007, *20*(2), inside front cover.

18. R.B. Blackburn and J. Payne, *Green Chem.*, 2004, *6*, G59.

19. (a) G.A. Davis, *The Use of Life Cycle Assessment in Environmental Labeling Programs*, EPA/742-R-93-003, Washington, DC, Sept. 1993; (b) M.A. Currran and T.J .Skone, *Environ. Prog.*, 2003, *22*(1), 1.

20. (a) B. Allen, *Green Chem.*, 2000, *2*, G19; (b) L.H. Gulbrandsen, *Environment*, 2005, *47*(5), 8.

21. (a) J.A. Dixon, L.F. Scura, R.A. Carpenter, and P.B. Sherman, *Economic Analysis of Environmental Impacts*, 2nd ed., Earthscan, London, 1994; (b) B.B. Marriott, *Environmental Impact Assessment: A Practical Guide*, McGraw-Hill, New York, 1997; (c) E.J. Calabrese and L.A. Baldwin, *Performing Ecological Risk Assessments*, St. Lucie Press, Delray Beach, FL, 1993; (d) P.A. Erickson, *A Practical Guide to Environmental Impact Assessment*, Academic, San Diego, CA, 1994; (e) S. Farrow and M. Toman, *Environment*, 1999, *41*(2), 12.

22. (a) D.J. Hanson, *Chem. Eng. News*, July 17, 1995, 45; (b) G.W. Suter, II, *Ecological Risk Assessment*, 2nd ed., CRC Press, Boca Raton, FL, 2007; (c) R.L. Revesz and M.A. Livermore, *Rethinking Rationality—How Cost-Benefit Analysis Can Better Protect the Environment and Our Health*, Oxford University Press, New York, 2008.

23. J. Johnson, *Chem. Eng. News*, Aug. 24, 1998, 14.

24. (a) L.R. Raber, *Chem. Eng. News*, Feb. 3, 1997, 28; (b) C.M. Cooney, *Environ. Sci. Technol.*, 1997, *31*, 14A; (c) T. Agres, *R&D (Cahners)*, 1998, *40*(5), 15.

25. (a) M. O'Brian, *Making Better Environmental Decisions: An Alternative to Risk Assessment*, MIT Press, Cambridge, MA, 2000; (b) Q. Zhang, J.C. Crittenden, and J.R. Mihelcic, *Environ. Sci. Technol.*, 2001, *35*, 1282; (c) H. Gavaghan, *Science*, 2000, *290*, 911; (d) F. Ackerman and L. Heinzerling, *Priceless—On Knowing the Price of Everything and the Value of Nothing*, The New Press, New York, 2004; (e) O. Pilkey, *Science*, 2008, *320*, 1423; (f) C. Hogue, *Chem. Eng. News*, Jan. 29, 2007, 32; (g) C.D. Brewer, *Science*, 2009, *325*, 1075.

26. P. Hawken, The Ecology of Commerce, A Declaration of Sustainability, Harper-Collins, New York, 1993, 177.

27. (a) J.A. Cichowicz, *How to Control Costs in Your Pollution Prevention Program*, Wiley, New York, 1997; (b) J.H. Clark, *Chemistry of Waste Minimisation*, Chapman & Hall, London, 1995; (c) S.T. Thomas, *Facility Manager's Guide to Pollution Prevention and Waste Minimization*, BNA Books, Washington, DC, 1995; (d) J.R. Aldrich, *Pollution Prevention Economics: Financial Impact on Business and Industry*, McGraw-Hill, New York, 1996; (e) R.C. Kirkwood and A.J. Longley, *Clean Technology and the Environment*, Blackie Academic, London, 1995; (f) P. Sharratt and M. Sparshott, *Case Studies in Environmental Technology*, IChemE Rugby, UK, 1996; (g) N.P. Cheremisinoff and A. Bendavid-Val, *Green Profits*, Butterworth-Heinemann, Woburn, MA, 2001; (h) J.D. Underwood, *Chem. Ind., (Lond.)*, Jan. 3, 1994, 18.

28. A. Thayer *Chem. Eng. News*, July 3, 1995, 10.

29. (a) *Green Ledgers: Case Studies in Corporate Environmental Accounting*, World Resources Institute, Washington, DC, 1995; (b) M. Spitzer and H. Elwood, *An Introduction to Environmental Accounting as a Business Management Tool: Key Concepts and Terms*, U.S. Environmental Protection Agency, EPA 742-R-95-001, June 1995.

30. D. Shannon, *Environ. Sci. Technol.*, 1995, *29*, 309A.

31. M. Dorfman, C. Miller, and W. Muir, *Environmental Dividends: Cutting More Chemical Wastes*, Inform, New York, 1992.

32. Anon., *Environ. Sci. Technol.*, 1994, *28*, 214A.

33. (a) A. Thayer, *Chem. Eng. News*, May 5, 1997, 28; (b) J. Johnson, *Chem. Eng. News*, Nov. 2, 2009, 24.

34. D. Munton, *Environment*, 1998, *40*(6), 4.

35. (a) S.L. Wilkinson, *Chem. Eng. News*, Aug. 4, 1997, 35; (b) Anon., *Green Chemistry and Engineering Conference*, Washington, DC, June 23–25, 1997; (c) P.V. Tebo., *Chemtech*, 1998, *28*(3), 8.

36. (a) K. Schmidt, *New Sci.*, 1996, *150*(2032), 32; (b) B. Hileman, *Chem. Eng. News*, May 29, 1995, 34; (c) S.M. Edgington, *Biotechnology*, 1995 *13*, 33.

37. H. Grann. In: D.J. Richards, ed., *The Industrial Green Game—Implications for Environmental Design and Management*, Royal Society of Chemistry, Cambridge, UK, 1997, 117–123.

38. D.J. Hanson, *Chem. Eng. News*, Dec. 8, 1997, 18.

39. J.D. Underwood, *Chem. Ind.*, (*Lond.*), 1994, 18.

40. E. Kirschner, *Chem. Eng. News*, Feb. 26, 1996, 19.

41. J. Johnson, *Chem. Eng. News*, Sept. 29, 1997, 19.

42. J.J. Romm, *Cool Companies: How the Best Businesses Boost Profits and Productivity by Cutting Greenhouse Gas Emissions*, Island Press, Washington, DC, 1999.

43. (a) P. Wilmshurst, *Chem. Ind.* (*Lond.*), 1997, 706; (b) M. Jacobs, *Chem. Eng. News*, May 17, 1999, 5; (c) S. Garrattini, *Science*, 1997, *275*, 287.

44. R.U. Ayres, *Environ. Sci. Technol.*, 1998, *32*, 366A.

45. (a) W. Greider, *On Earth (Natural Resources Defense Council)*, Fall, 2003, 20; (b) A.W. Savitz and K. Weber, *The Triple Bottom Line—How Today's Best-Run Companies Are Achieving Economic, Social and Environmental Success*, Jossey-Bass, New York, 2006.

46. (a) Anon., *Environ. Sci. Technol.*, 2000, *34*, 459A; (b) Anon., *Environment*, Nov., 2000, 6.

47. (a) J. Motavelli, *Environmental Defense Solutions*, 2004, *35*(1), 10; (b) V. Dunn, *Chem. Ind.* (*Lond.*), Feb. 2, 2004, 16.

48. (a) K. Elllison, *Nat. Conserv. Mag.*, 2002, *52*(4), 44; (b) www.innovestgroup.com/home.htm; (c) www.equator–principles.com/principles.shtml; (d) Environmental valuation and cost-benefit news, *Sustainable Asset Management Sustainability Yearbook 2008*, Apr. 7, 2008; (e) Anon., *Chem.Eng. News*, Sept. 28, 2009, 33.

49. (a) T. Whelan, *Nature Tourism—Managing for the Environment*, Island Press, Washington, DC, 1991; (b) G. Neale, ed., *The Green Travel Guide*, 2nd ed., Earthscan, London, 1999; (c) M. Honey, *Ecotourism and Sustainable Development—Who Owns Paradise?* Island Press, Washington, DC, 1998; (d) R.B. Primack, D. Bray, H.A. Galletti, and I. Ponciano, eds, *Timber, Tourists and Temples—Conservation and Development in the Maya Forest of Belize, Guatemala and Mexico*, Island Press, Washington, DC, 1998; (e) L. France, *The Earthscan Reader in Sustainable Tourism*, Earthscan, London, 1997; (f) H. Youth, *World Watch*, 2000, *13*(3), 120; (g) Anon., *Environment*, 1999, *41*(10), 7; (g) D.B. Weaver, ed., *The Encyclopedia of Ecotourism*, Oxford University Press, 2001; (h) G. Neale, ed., *The Green Travel Guide*, 2nd ed., Earthscan, London, 1999; (i) D.A. Fennell, *Ecotourism Program Planning*, Oxford University Press, UK, 2002.

50. *J. Ecotourism* and *J. Sustainable Tourism*, both from Channel View Publications, Bristol, UK.

51. (a) J. Tidwell, *Conservation Front Lines*, Conservation International, Washington, DC, 2006, *6.1*, 6; (b) P.J. Ferraro and A. Kiss, *Science*, 2002, *298*, 1718; (c) G.C. Daily, T. Soderquist, S. Aniyar, K. Arrow, P. Dasgupta, P.R. Ehrlich et al., *Science*, 2000, *289*, 395.

52. (a) R. Buckley, *Case Studies in Ecotourism*, Oxford University Press, UK, 2003; (b) R. Buckley, C. Pickering, and D.B. Weaver, eds, *Nature-Based Tourism, Environment and Land Management*, Oxford University Press, UK, 2004; (c) L. France, *The Earthscan Reader in Sustainable Tourism*, Stylus Publishing, Herndon, VA, 1997; (d) P.W. McRandle, *World Watch*, 2006, *19*(4), 5; (e) P.F.J. Eagles, S.F. McCool, and C.D. Haynes, *Sustainable Tourism in Protected Areas—Guidelines for Planning and Management*, Island Press, Washington, DC, 2002; (f) A. Ananthaswamy, *New Sci.*, Mar. 6, 2004, 6.

53. M. Honey, *Environment*, 2003, *45*(6), 8.

54. A.T. White, V. Barker, and G. Tantrigama, *Ambio*, 1997, *26*, 335.

55. S.C. Rai and R.C. Sundriyal, *Ambio*, 1997, *26*, 235.

56. D. Knigh, B. Mitchell, and G. Wall, *Ambio*, 1997, *26*, 90.

57. V.J. Harriott, D. Davis, and S.A. Banks, *Ambio*, 1997, *26*, 173.

58. (a) A. Esty, *Am. Sci.*, 2004, *92*(6), 513; (b) Anon., *Conserv. Mag.*, 2009, *10*(1), 22.

59. R. Socolow, C. Andrews, F. Berkhout, and V. Thomas, eds, *Industrial Ecology and Global Change*, Cambridge University Press, Cambridge, 1994, 406.

60. (a) J. Barrett, *Pulp Pap. Int.*, 1994, *36*(12), 53; (b) M. Wagner, *Amicus J. (Natural Resources Defense Council)*, 1997, *19*(1), 17; (c) *Amicus. J. (Natural Resources Defense Council)*, 1998, *20*(2), 46; (d) B. Allen, *Green Chem.*, 2000, *2*, G19.

61. J. Gersh, *Amicus. J. (Natural Resources Defense Council)*, 1999, *21*(1), 40.

62. N. Dudley, C. Elliott, and S. Stolton, *Environment*, 1997, *39*(6), 16.

63. M.S. Reisch, *Chem. Eng. News*, Jan. 12, 1998, 98.

64. (a) B. Hileman, *Chem. Eng. News*, Feb. 9, 1998, 16; (b) J.A. Tickner, ed., *Precaution: Environmental Science, and Preventive Public Policy*, Island Press, Washington, DC, 2002.

65. (a) D.D. Nelson, *International Environmental Auditing*, Government Institutes, Rockville, MD, 1998; (b) P. Sands, *Principles of International Environmental Law*, 2nd ed., Cambridge University Press, Cambridge, 2003; (c) P. Sands and P. Galizzi, eds, *Documents in International Environmental Law*, 2nd ed., Cambridge University Press, Cambridge, 2004.

66. (a) B. Hileman, *Chem. Eng. News*, June 23, 1997, 26; (b) E. Papadakis, *Environmental Politics and Institutional Change*, Cambridge University Press, New York, 1997.

67. Anon., *Chem. Ind. (Lond.)*, 1998, 376.

68. J.-F. Tremblay, *Chem. Eng. News*, May 31, 1999, 19.

69. M.G. Malloy, *Waste Age's Recycling Times*, 1998, *10*(6), 6.

70. (a) A. Gouldson and J. Murphy, *Regulatory Realities—the Implementation and Impact of Industrial Environmental Regulation*, Earthscan, London, 1998; (b) Anon., *Chem. Eng. News*, Nov. 9, 1998, 18; (c) J. Johnson, *Chem. Eng. News*, Jan. 24, 2000, 14.

71. R. Socolow, C. Andrews, F. Berkhout, and V. Thomas, eds, *Industrial Ecology and Global Change*, Cambridge University Press, Cambridge, 1994, 384.

72. (a) Anon., *Environ. Sci. Technol.*, 1998, *32*, 213A; (b) B. Hileman, *Chem. Eng. News*, Mar. 23, 1998, 30; (c) P.R. Portney, *Environment*, 1998, *40*(2), 14; (d) C. Hogue, *Chem. Eng. News*, Sept. 27, 2003, 27; Feb. 26, 2001, 32; Aug. 23, 2004, 26.

73. Anon., *Chem. Eng. News*, Dec. 1, 1997, 14.

74. Anon., *Chem. Eng. News*, Apr. 19, 1999, 38.

75. (a) F. Krupp, *EDF Lett.* (Environmental Defense Fund), 1998, *29*(1), 3; (b) G.T. Svendsen, *Public Choice and Environmental Regulation—Tradable Permit Systems in the United States and Carbon Dioxide Taxation in Europe*, Edward Elgar, Cheltenham, UK, 1998; (c) S. Sorrell and J. Skea, *Pollution for Sale: Emissions Trading and Joint Implementation*, Edward Elgar, Cheltenham, UK, 1999.

76. (a) R.A. Kerr and J. Kaiser, *Science*, 1998, *282*, 1024; (b) B. Hileman, *Chem. Eng. News*, Mar. 2, 1998, 28; (c) J. Boyd, D. Burstraw, A. Krupnick, V. McConnell, R.G. Newell, K. Palmer, J.N. Sanchiro, and M. Walls, *Environ. Sci. Technol.*, 2003, *37*, 216A.

77. (a) M. Webster, *On Earth*, 2009, *31*(1), 58; (b) G.T. Svendsen, *Public Choice and Environmental Regulation—Tradable Permit Systems in the United States and Carbon Dioxide Taxation in Europe*, Edgar Elger, Cheltenham, UK, 1998; (c) S. Sorrell and J. Shea, *Pollution for Sale: Emissions Trading and Joint Implementation*, Edgar Elger, Cheltenham, UK, 1999; (d) P.C. Fusaro and M. Yuen, *Green Trading Markets: Developing the Second Wave*, Elsevier Science, Amsterdam, 2005.

78. *New Jersey Discharger* (New Jersey Department of Environmental Protection), Trenton, NJ, 1997, *5*(1).

79. S. Schaltegger and T. Thomas, *Ecol. Econ.*, 1996, *19*, 35.

80. (a) J.B. Hockenstein, R.N. Stavins, and B.W. Whitehead, *Environment*, 1997, *19*(4), 13; (b) J.R. Aldrich, P*ollution Prevention Economics: Financial Impacts on Business and Industry*, McGraw-Hill, New York, 1996, 24; (c) *The Distributive Effects of Economic Instruments for Environmental Policy*, Organization for Economic Cooperation and Development, Paris, 1994; (d) *Managing the Environment: The Role of Economic Instruments*, Organization for Economic Cooperation and Development, Paris, 1994; (e) *Environmental Policy: How to Apply Economics Instruments*, Organization for Economic Cooperation and Development, Paris, 1991; (f) Anon., *Economic Instruments for Environmental Protection*, Organization for Economic Cooperation and Development, Paris, 1989; (g) Anon., *Renewable Natural Resources: Economic Incentives for Improved Management*, Organization for Economic Cooperation and Development, 1989; (h) Anon., *Saving Biological Diversity: Economic Incentive*, Organization for Economic Cooperation and Development, Paris, 1997.

81. R. Hueting, *Ecol. Econ.*, 1996, *18*, 81.

82. (a) Anon., *Chem. Eng. News*, Mar. 20, 1995, 17; (b) R.H. Bezdek, *Ambio*, 1989, *18*, 274; (c) M. Renner, Jobs in a sustainable economy, Worldwatch paper 104, Sept. 1991, Worldwatch Institute, Washington, DC; (d) Anon., *Futurist*, Mar.–Apr., 1992, *32*, 49; (e) Anon., *Chem. Eng. News*, Nov. 16, 1992, 12; (f) J. Johnson, *Environ. Sci. Technol.*, 1995, *29*, 19A; (g) R.H. Bezdek, *Environment*, 1993, *35*(7), 7; (h) E. Goldstein, *Environ. Manage.*, 1996, *20*, 313; (i) *Environ. Sci. Technol.*, 1995, *29*, 310A; (j) L. Ember, *Chem. Eng. News*, Jan. 23, 1995, 19; (k) E.S. Goodstein, *The Trade-Off Myth: Fact and Fiction about Jobs and the Environment*, Island Press, Washington, DC, 1999.

83. www.bottlebillhawaii.org/factls.htm.

84. Anon., *Chem. Eng. News*, Mar. 20, 1995, 17.

85. (a) Anon., *Environment*, 2000, *42*(5), 6; (b) Anon., *Chem. Eng. News*, Sept. 25, 2000, 33.

86. (a) Anon., *Environ. Sci. Technol.*, 1995, *29*, 173A; (b) *Chem. Ind. (Lond.)*, 1995, 361.

87. (a) J. Johnson, *Chem. Eng. News*, Dec. 1, 1997, 15; (b) D.R. Berg and G. Ferrier, *Chemtech*, 1999, *29*(3), 45; (c) Anon., *Chem. Eng. News*, June 22, 1998, 22; (d) K.S. Betts, *Environ. Sci. Technol.*, 1998, *32*, 353A.

88. Anon., *Chem. Ind. (Lond.)*, 1994, 398.

89. Anon., *Chem. Eng. News*, Nov. 2, 1998, 16.

90. (a) L. Jarvis, *Amicus. J. (Natural Resources Defense Council)*, 1998, *20*(2), 24; (b) J. Makower, *Good, Green Jobs*, California Department of Conservation, Sacramento, CA, 1995.

91. (a) W. Nixon, *Amicus. J. (Natural Resources Defense Council)*, 1998, *20*(2), 16; (b) M. Reisner, *Amicus. J. (Natural Resources Defense Council)*, 1998, *20*(2), 19.

92. (a) Anon., *Environ. Sci. Technol.*, 1997, *31*, 82A; (b) N. Myers and J. Kent, *Perverse Subsides: Tax $s Undercutting Our Economies and Environments Alike*, International Institute for Sustainable Development, Winnipeg, 1998; (c) J. Gersh, *Amicus. J. (Natural Resources Defense Council)*, 1999, *21*(1), 37; (d) D.M. Roodman, *World Watch*, 1995, *8*(5), 13; (e) N. Myers, *Science*, 2000, *287*, 2419; (f) N. Myers and J. Kent, *Perverse Subsidies—How Misused Tax Dollars Harm the Environment and the Economy*, Island Press, Washington, DC, 2001; (g) Worldwatch Institute, *State of the World 2008*,

W.W. Norton, New York, 2008; (h) A. Balmford, A. Bruner, P. Cooper, R. Costanza, S. Farber, R.E. Green M. Jenkins et al., *Science*, 2002, *297*, 950; (i) P.R. Ehrlich and A.H. Ehrlich, *One With Nineveh—Politics, Consumption and the Human Future*, Island Press, Washington, DC, 2004; (j) C. Pye-Smith, *The Subsidy Scandal—How Your Government Wastes Your Money to Wreck Your Environment*, Stylus Publishing, Herndon, VA, 2002.

93. (a) T. Prugh, *Natural Capital and Human Economic Survival*, ISEE Press, Solomons, MD, 1995, *114*, 140; (b) M. Khanna and D. Zilberman, *Ecol. Econ.*, 1997, *23*(1), 25.
94. E. Rodenburg, *Environment*, 1997, *39*(4), 25.
95. J.M. Heumann, *Waste Age's Recycling Times*, 1997, *9*(24), 2.
96. (a) B. Hileman, *Chem. Eng. News*, June 23, 1997, 26; (b) C. Lynch, *Amicus. J. (Natural Resources Defense Council)*, 1998, *19*(4), 15.
97. R. Gelbspan, *Amicus. J. (Natural Resources Defense Council)*, 1998, *19*(4), 22.
98. D. Pimentel, C. Wilson, C. McCullum, R. Huang, P. Dwen, J. Flack, Q. Tran, T. Saltman, and B. Cliff, *Bioscience*, 1997, *47*, 754.
99. C. Anderson, *Wilmington Delaware News J.*, Dec. 13, 1997, F5.
100. C. Coles, *Futurist*, 2003, *37*(3), 13.
101. (a) D. Pauly, V. Christensen, J. Dalsgaard, R. Froese, and F. Torres, Jr., *Science*, 1998, *279*, 860; (b) N. Williams, *Science*, 1998, *279*, 809; (c) C. Ash, *Science*, 2004, *305*, 1242.
102. (a) B.S. Dunkiel, *Environment*, 1996, *38*(10), 16; (b) M. Hyman, *Chem. Ind. (Lond.)*, 1999, 528; (c) Green Tax Database, http://www.oecd.org/env/policies/taxes/index.htm.
103. (a) *Taxation and the Environment*, Organization for Economic Cooperation and Development, Paris, 1993; (b) S. Bernow, R. Costanza, H. Daly, R. DeGennaro, D. Erlandson, D. Ferris, P. Hawken, J.A. Hoerner, J. Lancelot, T. Marx, D. Norland, I. Peters, D.L. Roodman, C. Schneider, P. Shyamsundar, and J. Woodwell, *BioScience*, 1998, *48*, 193; (c) T. O'Riordan, ed., *Ecotaxation*, Earthscan, London, 1997; (d) D.M. Roodman, *World Watch*, 1995, *8*(5), 13.
104. (a) Anon., *Environment and Taxation: The Cases of the Netherlands, Sweden, and the United States*, Organization for Economic Cooperation and Development, Paris, 1994; (b) M. Burke, *Environ. Sci. Technol.*, 1997, *31*, 84A; (c) Anon., *Chem. Ind. (Lond.)*, 1996, *38*(3), 16; (d) F. Muller, *Environment.*, 1996, *38*(2), 12; (e) D. Cansier and R. Krumm, *Ecol. Econ.*, 1997, *23*(1), 59; (f) C. Hanisch, *Environ. Sci. Technol.*, 1998, *32*, 540A; (g) P. Layman, *Chem. Eng. News*, June 14, 1999, 14.
105. L. Ember, *Chem. Eng. News*, June 2, 1997, 8.
106. R. Tyson, *Environ. Sci. Technol.*, 1997, *31*, 454A.
107. Anon., *Environ. Sci. Technol.*, 1997, *31*, 407A.
108. C. Hanisch, *Environ. Sci. Technol.*, 1998, *32*, 540A.
109. Anon., *Chem. Ind. (Lond.)*, 1998, 248.
110. C. Martin, *Chem. Br.*, 1997, *33*(12), 33.
111. R.A. Bohm, C. Ge, M. Russell, J. Wang, and J. Yang, *Environment*, 1998, *40*(7), 10.

6 The Greening of Society

6.1 INTRODUCTION

Chapter 5 described some economic and institutional factors that slow down the adoption of pollution prevention. This chapter continues this theme by considering the extent to which individuals, governments, and businesses have embraced the concepts of a clean environment and a sustainable future.

Gaylord Nelson, a former senator from Wisconsin, originated Earth Day, the first one being held in 1970.[1] This event and others contributed to the rise of the environmental movement in the United States. The United States Congress responded to this movement by enacting the Clean Air Act, the Clean Water Act, and the Endangered Species Act, as well as other laws. What has happened since then? One poll conducted in the United States in 1994 found that 23% of the respondents considered themselves to be active environmentalists, and 56% were sympathetic to the environmental cause.[2] More than half felt that environmental laws had not gone far enough, with only 16% feeling that the laws had gone too far. Environmental groups were considered favorably by 74% of the respondents.

Progress has been made in some areas since 1970. Emissions of pollutants from point sources into air and water have decreased. Toxic releases are decreasing. Some Superfund sites have been cleaned up. Businesses would no longer think of dumping a barrel of waste solvent on the ground at the landfill site so that the barrel could be used again for the same purpose. Control of pollutants from nonpoint sources is still a problem. There is now more international cooperation and discussion of global problems, such as ozone depletion by chlorofluorocarbons (CFCs) and the effect of population growth on the environment. Several nations have achieved zero population growth rates. Population growth rates are still high in some of the developing nations least able to deal with them. A great many of the world's fisheries are overfished. Biodiversity continues to decline. More consumer goods are marked with the postconsumer recycled content. By analyzing the greening that has occurred during this period, it may be possible to find ways to speed up the process of attacking the remaining problems.

6.2 INDIVIDUALS

Many persons who say that they are green may be so only in limited ways, owing to problems that they do not associate themselves with or do not understand.[3] Use the quiz below to see how green you are:

1. What do you put on your lawn?
2. What do you take off your lawn and what do you do with it?
3. What do you use to carry the groceries in after you buy them?

4. Do you use any single-use, throwaway items?
5. How many cars of what type are there in your family?
6. How long has it been since you rode on a train?
7. How do you dry clothes after washing?
8. How often do you mend an object instead of discarding it and buying a new one?
9. What do you do with clothes or toys that the children have outgrown?
10. Do you have anything in your home or office that was made using toxic chemicals?
11. Do you own anything made of teak or mahogany?
12. What do you recycle?
13. How many children do you have or intend to have?
14. When did you last contact your congressman?
15. Does your home or office have active or passive solar heating and cooling? Could these be added?
16. Does the wastewater from your toilet receive tertiary treatment?
17. How many suits or dresses do you own?
18. How far do you live from work?
19. Do you obey the speed limits on highways?

Runoff from lawns, golf courses, and farms contaminates many streams, decreasing the biodiversity in them. About 18% of municipal solid waste is yard waste. If you put grass clippings or leaves in the trash can, you are contributing to the landfill problem. Several of the questions are directed at using less material and less energy. Groceries can be carried in string bags, or in your own basket. There are many times where you can substitute reusable items for single-use ones. Mending something can substitute for buying a new one. Passing things that you no longer need on to friends can eliminate their need to buy something. Living close to where you work may allow you to get there by walking or cycling. Taking the train instead of your car or an airplane can save energy. So can drying washing on an outdoor line. There is nothing wrong with changing your own oil in your car. The problem is that about half of it ends up going down storm drains to contaminate streams.

You may have a surprising number of chemicals considered to be hazardous in your home. These include pesticides, lye for cleaning drains, hydrocarbon solvents for cleaning, paints, varnishes, and others. Check the list of chemicals that household hazardous waste drives try to collect. Figuring out what you own that was made with toxic chemicals will be more of a job. You can probably figure out what formaldehyde was used to make, but beyond that it may be difficult. Do not forget that the gasoline in your car was probably made using an alkylation step catalyzed by hydrogen fluoride or sulfuric acid. The leather in your shoes was probably tanned with a chromium salt. For the person with a chemical background, a search of industrial encyclopedias and books on industrial chemistry may help you decide what toxic chemicals were used. It is also true that some toxic chemicals are so useful that it will be a long time before they are displaced. Acrylonitrile and ethylene oxide fall in this class. If the company that made the item that you are using has a consumer service telephone number, you can ask them what hazardous chemicals were used in its manufacture. They may or may not tell you.

Consumer choices can make a difference.[4] Energy efficiency could be increased if consumers heeded the energy efficiency rating on new cars and appliances. Energy would be saved and tailpipe emissions reduced if they bought small cars instead of vans, sport utility wagons, and pickup trucks. Wood is a renewable form of energy, but if a person buys and uses a wood stove without a catalytic combustion aid in the stack, the stove will be a source of air pollution. It is difficult for consumers to test different brands on their own. However, they may be able to obtain the information needed to make an intelligent decision from a consumer-testing group such as Consumer Reports. It takes effort to be really green. It may mean more time in the library and more time shopping around for a reusable item instead of the throwaway version that is more common in the stores. It also takes extra effort to read the green logo that the manufacturer has put on a product to determine that it is really meaningful and that important points, which might be negative, have not been left out. Green consumerism is said to be making a difference and is pushing manufacturers into offering products that are friendlier to the environment.[5]

6.3 GOVERNMENT

Governments are supposed to take the long view that will do the most good for the greatest number of people in the long term. Ecopolitics leading to a green, sustainable society should be part of this.[6] This trend seemed to be true from 1970 up to 1995 with the passage of numerous environmental laws in the United States. The air and water of the country became cleaner during this period. Then the anti-environmental backlash hit as the 104th Congress convened in 1995. This proved to be the worst Congress on environmental matters in 25 years, as rated by the League of Conservation Voters.[7] With a maximum score of 100, 111 representatives and 24 senators received zeros. It was an environmental backslide[8] as efforts were made to weaken major environmental laws and cut funding for renewable energy studies, the EPA, national parks, and others in the name of "regulatory reform," removal of "burdensome regulations," and "balancing the budget."[9] A number of the attempts showed up as riders on unrelated appropriation bills, presumably to avoid floor debate and with the hope that they would pass because other congressmen would not want to delay an important appropriations bill. This happened despite the overwhelming support of the public for strong environmental and public health standards. That the efforts failed in large part was probably due to this public, which made its feelings known to the Congress and to the administration. The administration included the staunch environmentalist Vice President Al Gore.[10] The Office of Technology Assessment, which won praise from many people for its reports over 23 years, was a casualty of the 104th Congress.[11] Many environmentalists would like to see it started again.

The American public had greened over the years, but many of the special interest groups had not. Part of the problem with the 104th Congress was the power of campaign contributions. It takes a lot of money to run for election to Congress. Congressmen had voted three out of four times for bills (to relax federal environmental protection) that were supported by groups that had given them money through political action committees.[12] These groups included agribusiness, chemical manufacturers, and natural resource extraction companies. Many chemical companies, as well as the American Chemistry Council, also give "soft" money to the major political parties.[13] Many congressmen were still not green in 1999. The U.S. House of Representatives passed the Mandates Information Act (H.R. 350), which requires estimates by the Congressional Budget Office of the costs imposed on the private sector by a regulation, but requires no estimates of its benefits.[14]

Europe has greened as well over the years. (There are green parties in at least 13 countries in Europe.)[15] The Green Party in Germany has been active. One result is the system making manufacturers responsible for the disposal of what they produce at the end of its useful life. Several countries in Europe are incorporating ecotaxes into their tax systems. Thirteen countries in Europe have achieved zero population growth. Most companies in the United Kingdom do not find environmental laws to be onerous.[16] The EU has greened more in some ways than the United States, so that there have been conflicts over climate change, genetically engineered crops, and the use of hormones in raising beef.[17]

Under the guidance of the United Nations, more and more attention is being given to the earth's environmental problems,[18] including that of environmentally sustainable development. Many international conferences sponsored by the United Nations have dealt with world problems. The nations of the world have frequently recognized the problems. Actions taken (Montreal Protocol of 1987) have been effective in the phase-out of the CFCs that threaten the ozone layer.[19] The 1997 conference, in Kyoto, Japan, made a start in mitigating global warming (i.e., if the nations can actually implement the reductions of greenhouse gases agreed upon). In other instances, the promise of improvement remains largely unfulfilled.[20] Ten years after the Brundtland report[21] on the need for a sustainable future and 5 years after the Earth Summit at Rio de Janeiro,[22] most of the problems identified were still there in 1997 or had worsened. The same was true after the meeting in Johannesburg in 2002.[23] (There have also been some problems in persuading some countries to comply with the treaties, such as the convention on International Trade in Endangered Species and the ban on CFCs.)[24] The United States has tried to eliminate a black market in CFCs.[25]

Nongovernmental organizations (NGOs) can send observers, but not delegates, to these international conferences. Despite this second-rate status, they have influenced the process of sustainable development.[26] While at conferences, they can disseminate information and offer new ideas and suggestions. The idea of sustainable development has become part of mainstream thinking, much more so than 25 years ago.

6.4 BUSINESSES

Many businesses have become greener since the founding of the U.S. EPA in 1970[27] and some more than others.[28] R. E. Chandler of Monsanto describes the response of the chemical industry to environmentalism as consisting of three overlapping stages.[29] The first phase was the denial phase. Industry tended to deny that major problems existed. It often predicted dramatically higher costs, job losses, and plant closures if the regulations were enforced. This exaggeration caused it to lose credibility with both legislators and the general public. The second phase was the "risk–benefit" one. Major U.S. chemical companies launched expensive public relations campaigns to convince the public that its emotions and fears of the unknown should be replaced by a more industry-like view of facts and reason in which the risk was small. Monsanto's Chemical Facts of Life of the late 1970s and early 1980s was such a program. The third phase involves the "public right to know," which may have been a reaction to the disaster at Bhopal, India. It includes a proactive openness with the community. The goal is to go beyond compliance with all laws to reduce all toxic and hazardous emissions, ultimately to zero. This phase incorporates pollution prevention. Some companies view environmental concerns as profit-drivers (i.e., through higher sales of cleaner products). The actions of the 104th Congress in the United States suggest that there are many companies that are still in the first phase.

Another author describes the first phase as an inherently defensive one, where management believes that it is being treated unjustly by the regulatory authority.[30] The third stage is one during which there is a sustainable, interactive program of environmental management within the company. Another author classifies corporations as red, yellow, or green, depending on the extent of their conversion to a long-term, opportunity-seeking, company-wide, cradle-to-grave performance system, which is the green one.[31] He and others feel that the green phase should offer a competitive advantage[32] and be part of a sustainable future.[33]

It is possible that the companies that grumble the most about environmental regulations are still in the first phase. It is said that, "No major piece of environmental legislation has ever been supported by corporate America."[34] A few examples will illustrate the complaining often coming from companies, or the trade association that represents them. The extent of the complaining may go up as the cost of compliance becomes high, or as the amount of change required to comply becomes large. Earnest W. Deavenport, Jr. of Eastman Chemical and the American Chemistry Council has pushed for regulatory reform.[35] An analysis sponsored by Pfizer concluded that the laws disregard reality, that the regulatory framework is impossible to administer properly, and that they foster excessive litigation.[36] The U.S. EPA, which administers the laws enacted by the U.S. Congress, is subjected to a lot of criticism. Its expansion of the number of chemicals to be reported in the Toxic Release Inventory was objected to by the National Council of Chemical Distributors (which has 333 members).[37] The American Petroleum Institute, representing refinery operators, says the cost of the new rule for reducing the emission of toxic chemicals from refineries will exceed the benefit.[38] The reduction would be 227,000 t annually. It also objected to the reduction of sulfur in gasoline.[39] The National Association of Manufacturers says that "There are some serious problems with our environmental regulatory system."[40] The association did find that over half of the members that voluntarily changed manufacturing processes to reduce waste or emissions saved money as a result. Industrial groups have fought the "credible evidence rule," which says that any evidence deemed credible can be used to determine whether a facility is violating emissions standards.[41] Electrical utilities have sued the EPA over a rule that would reduce the emission of nitrogen oxides by 900,000 t/yr.[42]

It is important to remember where regulations come from. Many are the result of tragedies, such as the Toxic Substances Control Act, which resulted from the accident at Bhopal, India. Others try to correct abuses that impair the health and safety of citizens that individuals and companies are not taking care of voluntarily. Speed limits on highways are intended to lower accidents, but many motorists exceed them. The Motiva (now Valero) refinery in Delaware deferred maintenance on a tank. The managers instructed operators not to fill the tank above the entry hole. A welder on the tank ignited the hydrocarbons in the sulfuric acid in the tank, which resulted in a fire and explosion that killed the welder and injured seven others, and spilled 1 million gallons of concentrated sulfuric acid. Now Delaware has a law that all tanks of 25,000 gallons or more must be registered and inspected by the state. The paperwork that goes with regulations is burdensome, but no one has found a good substitute for it.

Many U.S. industries are becoming greener based on what they say at meetings and in print.[43] However, it is not possible to tell how green they are by reading their annual environmental reports.[44] The level of disclosure is said to vary greatly. There is a high degree of subjectivity and selectivity in what is reported, often with a lack of quantification of environmental impacts. A survey of 54 large U.S. corporations, which expressed environmental concern through such reports, found that only one-third of the companies would switch to a less toxic material if it added 1% to the cost of the product, but only two said they would be willing to raise the cost by 5%.[45] Public relations firms can be hired to write such reports, as well as prepare advertisements, and even provide "citizen" letter-writing campaigns.[46] The advertisements may picture green forests and wildlife next to the plant. At the same time, the company, through its membership in a trade association, may be lobbying Congress to relax the laws that regulate it. One report says that businesses in Great Britain are paying little more than lip service to the environment.[47] Skeptics have been harsh in their criticisms of these practices.[48] They accuse firms of trying to spread doubt and uncertainty.

The CMA (now the American Chemistry Council) was 125 years old in 1997. Its over 190 member companies make over 90% of the chemicals that are produced in the United States.[49] Its future involves embracing sustainable development, establishing a health and environmental research program, and bringing greater flexibility into government regulations.[50] It is collaborating with the U.S. EPA in the testing of high-volume chemicals for toxicity and as endocrine disruptors.[51] (The Society of Chemical Manufacturers and Affiliates and the American Chemistry Council objects to the "design" of the program, fearing that the cost of testing will be a hardship on its members.[52] It also criticizes the U.S. EPA for ignoring the "unique needs of small-batch and custom chemical producers" and "its opposition to promising suggestions.")[53] The Chemical Industry Institute of Toxicology is funded by the American Chemistry Council.[54] The Council is proud of the record of the chemical industry in reducing releases of chemicals on the Toxics Release Inventory by more than 60% in six years.[55] However, it opposed expansion of the Toxics Release Inventory in the courts, losing its suit in the appeals court.[56] Its request to have ethylene glycol removed from the Inventory was denied by the EPA.[57] It does not support additional testing of the health effects of biphenyl, carbonyl sulfide, chlorobenzene, ethylene glycol, methyl isobutyl ketone, phenol, and trichlorobenzene, as suggested by the EPA.[58] It also opposes mandatory reporting of toxic chemical use, although such systems appear to be working well in Massachusetts and New Jersey.[59] The Synthetic Organic Chemical Manufacturers Association also objects to this.[60] Such inventory reporting should offer the maximum of flexibility for those companies who dislike "command-and-control" regulation. The Toxics Release Inventory requires no reduction in emissions at all, but companies often choose to make reductions after they realize how much they are losing in material and dollars.[61] They may also be motivated by wanting a good public image. The American Chemistry Council has spent US$40 million over 5 years on an advertising campaign to create a better public image for the chemical industry, but has discontinued the campaign as a result of its failure to change public attitudes.[62] The advertising campaign started again in 2005.[63]

Drug companies have sometimes filed frivolous patent applications to try to extend patent coverage on drugs. The companies claim that they need to maintain good profits to cover the high costs of research and development. This bothers the public when it finds out that the companies have been spending two to three times as much money on drug salesmen as on research and development.[64]

The showpiece of the American Chemistry Council is its "Responsible Care" program.[65] The program originated in Canada in 1985 and was adopted by the American Chemistry Council in 1988. It has spread to South America, Europe, Africa, Asia,[66] and Australia, and is now in 40 countries. The British Chemical Distributors and Traders Association is encouraging all of its members to sign up for the program.[67] Only 92 of its 117 members had done so by 1998. Companies pledge to adhere to 10 guiding principles:

1. Recognize and respond to community concerns.
2. Develop chemicals that are safe to make, transport, use, and dispose of.
3. Give priorities to environmental, health, and safety in planning products and processes.
4. Give information on chemical-related health and safety hazards promptly to officials, workers, and the public, and recommend protective measures.
5. Advise customers on the safe use and handling of products.
6. Operate plants in a way that protects the environment and the health and safety of workers and the public.
7. Conduct research on environmental, health, and safety aspects of products, processes, and waste.
8. Resolve problems created by the past handling and disposal of hazardous materials.
9. Participate with government and others to create responsible regulations to safeguard the community, workplace, and the environment.
10. Offer assistance to others who produce, use, transport, and dispose of chemicals.

Skeptics, who felt at first that the program was just a public relations ploy, now consider it to be real and worthwhile. They do point out that it may be written too broadly and that it may not be quantitative enough. The president of the American Chemistry Council has said, "We are acutely aware that the initiative loses credibility if we say one thing with Responsible Care and say another thing when we deal with Congress or the regulatory agencies." Independent third-party verification is starting to be used by some companies.[68] The United Nations Commission on Sustainable Development plans to "examine voluntary initiatives and agreements," but the American Chemistry Council does not want "a formal U.N. mechanism to oversee these programs."[69] Efforts need to be increased to bring labor into the program, for it has felt left out.[70] If this approach and similar ones prove workable and verifiable, this may be a good alternative to enactment of further regulations.[71]

Europe adopted a Registration, Evaluation, Authorisation and Restriction of Chemicals (REACH) law in 2008.[72] The American Chemistry Council believes that this law will stifle chemical innovation.[73] However, American companies that export chemicals to Europe will have to comply with it.

The American Chemistry Council was instrumental in having the comments of a respected toxicologist on polybrominated diphenyl ethers removed from an EPA assessment.[74] A regional EPA administrator was forced to resign for wanting Dow to remove the dioxins from two rivers in Michigan.[75] These events have led to a U.S. Congressional investigation of the influence of the American Chemistry Council on the EPA.[76] A bill to measure synthetic chemicals in the bodies of Californians, supported by health and environmental groups, but opposed by the American Chemistry Council, was vetoed by the governor.[77]

In 1988, the Coalition for Environmentally Responsible Economics (a group of investors and environmentalists) put together a code (first called the Valdez Principles, now the CERES Principles) for corporations to follow.[78] It includes the following:

1. Reducing releases into the biosphere and protecting biodiversity.
2. Sustainable use of natural resources.
3. Minimizing waste and disposing of it properly.
4. Sustainable use of energy.
5. Minimizing environmental, health, and safety risks to employees and the public.
6. Elimination of products that cause environmental damage.
7. Correcting any damage that may have been caused to people and to the environment.
8. Informing those involved of any potential hazards in a timely fashion.
9. Environmental interests must be represented on the Board of Directors.
10. An annual self-audit will be made public.

This code is broader than Responsible Care. It includes the concept of a sustainable future and requires that a self-audit be made public. Large companies have been slow to embrace it. Management usually recommends a vote against the many shareholder resolutions that propose it in the proxies for the annual meetings (e.g., GE in 1993 and Goodyear Tire and Rubber in 1996). Those that have adopted it include Sun Company, General Motors,[79] H. B. Fuller, Polaroid, and Arizona Public Service.[80]

The International Organization for Standardization (ISO) was formed in 1946 to facilitate standardization as a means of promoting international trade. It is located in

Geneva, Switzerland. Its ISO 9000 series, which deals with product quality, has become almost a requirement for companies to do business.[81] The series has been extended to an ISO 14000 series dealing with environmental management.[82] These standards include formalization of corporate policies and audits, evaluation of performance, labeling, and life-cycle assessments. The standard requires mandatory third-party verification. The goal is to support environmental protection in balance with socioeconomic needs. If these standards are adopted as widely as the ISO 9000 series has been, they may become a powerful force in cleaning up the environment. Companies buying the chemicals may insist that their suppliers adhere to the principles. This may be especially helpful in nations that lack an effective regulatory system.[83]

It is not easy to tell just how green a company is.[84] Stock analysts do not consider environmental performance a major factor in evaluating companies.[85] The U.S. National Academy of Engineering is developing methods to do this.[86] The Investor Responsibility Research Center evaluates companies based on Superfund sites, spills, total emissions, enforcement actions, and penalties.[87] (Such data can usually be obtained from state environmental departments in the United States.) The Hamburg Environmental Institute evaluated the environmental friendliness of nearly 70 of the largest chemical and pharmaceutical corporations in its annual Top 50 report.[88] The top 10 were deemed "proactive." The rankings of the seven U.S. corporations in this category in the report for 1995 were Johnson & Johnson first, 3M third, Procter & Gamble fourth, Dow fifth, Baxter International sixth, Bristol Meyers Squibb ninth, and DuPont tenth. Henkel (Germany) was second, Ciba–Geigy (Switzerland) seventh, and Unilever (Netherlands) eighth. Some of the rankings change with the year. DuPont was 27th in 1994. Unilever was 22nd in 1994. Six U.S. companies were among the 22 classified as "active," the next lower level. The remaining 18 companies were classified as "reactive," the next lower level, or "passive," a still lower level. Monsanto was 19th in the report. The U.S. firms classified as "passive" were Merck 43rd, Colgate–Palmolive 47th, Occidental 48th, and GE Plastics 49th. The lowest ranking, "negative," was given to firms that could not be ranked "due to insufficient communication of their environmental performance." Firms in this category included American Home Products, Exxon Chemical, Pfizer, Schering Plough, Warner–Lambert, and Rhone–Poulenc. Overall, companies were weakest when judged on the "sustainability" of their products.

The Council on Economic Priorities (a New York NGO) produces an annual list of the worst polluters in the United States.[89] These ratings are based on toxic releases, regulatory compliance, waste, cleanups, and worker health and safety of more than 100 companies. The 1994 list included Union Carbide (now Dow) (which had three times as many spills as the next worst competitor, as well as a higher than average frequency of health and safety violation), Exxon (now Exxon-Mobil), Texaco (now Chevron Texaco), International Paper, Southern Company, Westinghouse, Westvaco, and Maxxam (unsustainable harvesting of redwood forests). The 1995 list included Formosa Plastics, Exxon (now Exxon-Mobil), Maxxam, and Southern Company (the industry's top emitter of carbon dioxide). Texaco (now Chevron-Texaco), Union Carbide, Westinghouse, International Paper, and Westvaco were not on the 1995 list. "Businesses on the list call it biased, distorted or just plain wrong." Along with the list, the Council recommends ways in which the companies can improve.

At the other extreme is 3M, which has saved US$1.4 billion in its Pollution Prevention Pays campaign through 3450 projects during 1975–1991.[90] Pioneer Hi-Bred International developed a superior soybean for use as an animal food by inserting a gene from a Brazil nut.[91] When company testing showed that humans allergic to Brazil nuts would also be allergic to the soybean, it decided not to release the new plant. Even though the soybean was meant to be fed to animals, there was a possibility of its getting into food for humans.

Many companies appear to be somewhere between the extremes of "proactive" and "reactive," but the general direction is toward becoming greener. In addition to the criteria for judging companies already mentioned, there are some for which the necessary information is difficult to obtain. For example, it is hard to know what role a company plays in an anti-environmental stand taken by a trade association that it belongs to. Companies have also been criticized for being greener in things that are relatively inexpensive than in ones involving a lot of money, but that may have more environmental impact. There is also a legitimate debate about whether or not some company actions are green or not. For example, a new insecticide that contaminated air and water less than its predecessors would be an improvement, but not as much as less pesticide use through integrated pest management. The records of a few companies will be given to illustrate the range of greening. Some of these can be compared with the ratings given in the foregoing by environmental organizations.

Chevron had two oil spills and a release of silica–alumina catalyst in 1991–1993.[92] Its oil spills have been reduced by 97% since 1989. Losses owing to fire in 1993 were US$37 million. There was a fire and explosion in the hydrocracking unit in Richmond, California, in 1999.[93] Fines of US$1.4 million were paid in 1993. A fine of US$7 million was levied for leaks at its El Segundo marine terminal under the Clean Air Act.[94] One week later, the state of California fined it US$2 million for a leak in a pipeline carrying jet fuel.[95] The company is adding double-hulled tankers to its fleet as a way to reduce the possibility of oil spills.[96] The company has been on trial in Ecuador for the many oil spills in the Amazon that have not been cleaned up.[97] On the other hand, Jared Diamond has commended the company for its oil field operations in Papua New Guinea in a rain forest.[98] From 1989 to 1994, it had contributed about US$7 million "to hundreds of conservation, wildlife preservation and environmental research and educational organizations." The company has won awards for its work with wildlife habitat in Wyoming, Colorado, and other places.[99] Environmental groups have accused it of green-washing in its "People Do" and other advertising.[100] They claim that the company is frequently doing only what the law requires, while at the same time its lobbyists (through its membership in the American Petroleum Institute) are trying to convince the Congress to weaken the laws. Chevron's chairman has called for regulatory reform because, "Our regulatory system does not have a principle that says 'enough is enough.'" He says that the regulatory process has grown "bloated on an unbalanced diet of zero-risk thinking."[101] Company management recommended that shareholders vote against shareholder proposals at the annual meeting asking that the company not drill for oil in the Arctic National Wildlife Refuge, that the company publish a report on the environmental and safety hazards for communities surrounding its plants, and that the company publish a report on the contribution of its products to global warming, as well as efforts to moderate this.[102]

When Delaware's bottle bill and Coastal Zone Management Act were proposed in the 1970s, the DuPont Company testified against them in public hearings. Since then, the company has become greener. Paul Tebo of DuPont points to the company's Chattanooga, Tennessee, nylon plant, which now achieves a 99.8% yield of saleable products, a Lycra process that recovers 99.95% of its spinning solvent, a biodegradable oil for two-cycle engines, shipment of neoprene rubber in a Surlyn bag that can be compounded right into the rubber, and an herbicide plant in Java that has close to zero emissions.[103] According to him, "We're saying the reason we're making all these changes is not just that it is good for the environment, it's good for business." He is against the implementation of the Kyoto Protocol.[104] The company is striving for zero emissions. DuPont now makes methyl isocyanate from N-methylformamide on demand on site and no longer ships it. It also makes phosgene as needed, so that the amount on site is small. This is described as an interim measure until a better phosgene-free route to isocyanates can be developed. The company has now affiliated with the Pew Center on Global Climate Change, which supports the Kyoto Protocol for the reduction of greenhouse gases.[105]

Monsanto (which split into Monsanto and Solutia in 1997) has devised ways to make isocyanates without the use of phosgene. It has also found a way to make p-phenylenediamine antioxidants without the use of chlorine. The company pledge of 1990 includes the reduction of all toxic and hazardous releases and emissions, working toward the ultimate goal of zero effect.[106] It pledges also to "work to achieve sustainable agriculture through new technology and practices." It has offered two 1-million-dollar challenges to anyone who can help it find a cost-effective and commercially practical way to recover useful chemicals from its wastewater, at the same time reducing their effect on the environment. Both have been won by SRI International.[107] Monsanto also won a Presidential Green Chemistry Challenge Award in 1996 for a new method to make disodium iminodiacetate, an intermediate in the synthesis of the herbicide glyphosate.[108] The intermediate is made by the copper-catalyzed dehydrogenation of diethanolamine. The new route eliminates the use of hydrogen cyanide and formaldehyde, as well as eliminating 1 kg of waste for every 7-kg product. This "zero-waste" route gives a higher overall yield with fewer process steps. The company has produced Alachlor, an herbicide often found as a contaminant in groundwater, since 1969.[109] The company also markets insect-resistant corn and cotton, which incorporate genes for *Bacillus thuringiensis* toxins, as well as cotton and soybeans that are resistant to its herbicide glyphosate.[110] Environmentalists are concerned about the rate of buildup of resistance that will take place in insects and weeds. There is also the question of the extent of soil erosion with the soil kept bare except for the crop. Alternative agriculture could be a more sustainable alternative.

PROBLEMS

6.1   Manufacturing Products for a Sustainable Future

You work for a traditional chemical company that has not come forth with a really new product in many years. The new chief executive officer is a man of action. He is aware of the unsustainable course of the world's industrial nations and the current overconsumption of material goods.

In a bold move, he has told the financial analysts on Wall Street that he plans to switch the company to manufacture products for a sustainable future within two years. He expects to make more money by selling less material. This talk has sent the stock price up five points in one day. The public relations department employees are already at work on this proposal. They are looking for a new name for the company, for example, "Sustainability Unlimited," or perhaps "Innovation Plus." They are considering slogans that will market the new image, for example, "The future is now. Make more money with Innovation Plus." Your job is to make the announced policy come true. How can you possibly do this?

6.2 Remedying a Herbal Problem

You are the chief executive officer of a company that manufactures herbal supplements, which is part of an industry that does US$14 billion of business each year. Profits are very good, as about 20% of Americans take herbal supplements. You operate under the 1994 Supplements Act that you lobbied hard to get the United States Congress to enact. However, medical doctors and prescription drug companies are aggressively attacking your industry. They say that many herbal supplements are ineffective, unsafe, and vary widely in the amount of active agent present. In addition, claims are made that some may contain heavy metals or added prescription drugs, and the doses are not standardized. What can your industry do to fend off the attack?

6.3 Strong Paper Additives?

You are chemists employed in the research laboratories of Paper Chemicals, Incorporated. The company's dry strength additive ("Cruddy") was discovered in the 1950s. When added to paper at 5 lbs./t, the resulting product is twice as strong. After a slow start, "Cruddy" grew steadily until it now has 80% of the market with a profit margin of 40%. Company management has regarded it as a mature product and used it as a "cash cow." No scouting for new dry strength additives has been done in the past 20 years. The company's customer service representatives are on a first-name basis with all the mill superintendents and even allow them to win at golf tournaments. Recently, a competing product ("Strengthon") has entered the market. It gives twice the strength improvement at half the level of "Cruddy." Moreover, it does not have to be made into a solution at the mill with pH adjustment. Both products are marketed at the same price. Two of your biggest accounts have just switched to it and others are threatening to do so. What should you do?

6.4 A Strange New Disease

The United Nations Children's Fund wanted to reduce the incidence of sickness and death from waterborne diseases such as typhoid fever, dysentery, and cholera in Bangladesh and West Bengal, India. They succeeded in doing this by installing thousands of tube wells that were about 5 cm in diameter, and no more than 200 m deep. Wells had been installed for 80% of the population by the year 2000. Unfortunately, after a few years an epidemic of a strange new disease causing scaly skin lesions on the soles of the feet, hands, chest, arms, and legs appeared on thousands of people. There was also an increase in the number of internal cancers. How might you discover what the disease was, and if it was linked to the tube wells? How could the epidemic be banished?

6.5 "Adhere" versus "SuperGloop"

"Adhere" has been the industry standard carpet backing adhesive for many years. Last year it contributed 10% of your company's earnings. Since the carpet market is nearly saturated, and your company has most of the market share for adhesives used on it, no research has been done on this adhesive in recent times. The profits are now going into studies of the application of nanotechnology to carpet facing.

However, in recent months this pretty picture has started to fade. Another company has introduced a new adhesive ("SuperGloop") for the market currently dominated by "Adhere." It is 10% cheaper per pound, is lighter in color and only half as much is required. Customer loyalty is not working in the face of this superior product. Even though your salesmen have pointed out that the long-term durability hasn't been determined, many of your former customers are changing to the new adhesive. If you knew what "SuperGloop" is you might be able to produce a comparable product, and get back some of the business that you have lost. Should you take chemists off the nanotechnology project to start a crash program on a new adhesive? The alternative might be to sell the adhesive business and concentrate on more profitable core options. What should you do?

6.6 A Water Catastrophe

You are the chief engineer of a waterworks in a North American city. This is a pleasant land where cows dominate the rural landscape. The city has provided clean water to its residents for many years by using chlorine as a disinfectant. Suddenly, disaster has struck the city's residents. One hundred thousand people have become ill and several have died. What has happened? What has gone wrong, and how can it be fixed? Will you be able to keep your job or be tried in court for negligence?

6.7 Avoiding an Expensive Recoat

Your company has formulated a coating that can be used to protect buildings and equipment for life. This is the result of considerable research over a period of years. Your package of patents appears to be good. This development has enormous potential and should appeal to customers who wish to avoid the expense and inconvenience of recoating their items every few years or even more often. Sales have increased gradually and have reached over US$1 million a year. They appear about to take off, except for the effect of the depressed economy. In a couple more years, the research and development costs should be paid off and it should start making a profit.

However, a small company has begun offering a competing product that appears to infringe your patents. It is certainly inferior, but significantly cheaper. It is being produced by some former employees who left your company and starting making it in a converted supermarket in a decadent shopping mall. You are thinking of bringing an infringement suit against them, although the last time you hired courtroom attorneys it cost you a lot of money, almost all of a year's profit. The competing firm might not be able to afford legal help and the trial might be short. Would they try to invalidate your patents? Would they claim an improvement and demand cross licensing? Can you sue the former employees for theft of trade secrets? Is there a chance that they have really found something different? What should your company do?

6.8   Battling the Business Center

You are a manager at the research center of the EcoFresh Company ("where science leads the world"). Your company's process for manufacturing phenol from cumene is the dominant one in North America. However, EcoFresh has been reluctant to build a plant using the process due to the necessary capital cost. Lately, the company's business center has been saying that it should not compete with the many licensees of the process. One chemist is kept busy tweaking the synthesis to increase yields, and to troubleshoot the plants of the many licensees. The company has studied the preparation of bisphenol using HCl as the catalyst, but has been unable to do as well as the examples in the patents of other companies.

EcoFresh does manufacture dicumyl peroxide, for which the profit margin is good and a consistent earner. The company dominates the market for it as a cross-linking agent for black polyethylene wire coating (its major use)—a market that is not growing much. It brings in a few million U.S. dollars a year, which is enough to cover the CEO's bonus. You would like to assign a marketing specialist to grow the company by finding new applications, but the business center won't put up the money. A year's supply of dicumyl peroxide can be made in the plant in three months. You need to find other products to use the plant for the other nine months. Again, the business center does not want to spend any money. The management says it wants to get out of the cyclical commodity market and substitute more specialty products. It may be thinking of US$100 million/year in specialties.

Recently, a new threat has appeared. A Japanese company is offering colorless dicumylperoxide at a lower price that might be manufactured in a Chinese plant. Is this just an introductory price that they expect to raise later? Even though the brown color of your product doesn't make any difference in a black wire coating, some customers are switching suppliers. Should you start a crash program to remove the color? Must EcoFresh lower its price to meet the competition? Could you find a toll manufacturer in China to lower the cost of production? Should the business be sold to a smaller company with less overhead? Would you or the plant manager be designated as the person to find a buyer? Would one of your customers consider buying the business to be sure of having at least two sources of supply? What should you do?

6.9   Your Viewpoint on Some Green Chemistry and Sustainability Issues

1. Is nuclear energy the answer to global warming?
2. Are "natural" consumer products better options?
3. Have you used herbal supplements?
4. Is the FDA a friend or an impediment to progress?
5. Are we overregulated in society?
6. Are the laboratories of your university safer than those in the local chemical industry?

7. Is a well-tanned skin attractive?
8. Do you expect to have accidents as a chemist?
9. The local oil refinery has been closed. Is this good or bad? Many jobs have been lost. Should it clean up its act and reopen?
10. Do safety inspections do any good? Have you ever done any?
11. Do you like pretzels, peanuts, or potato chips with salt?
12. Are cell phone towers eyesores?

REFERENCES

1. (a) B. Ruben, *Environ. Action*, 1995, *27*(1), 11; (b) M. Jacobs, *Chem. Eng. News*, Apr. 24, 2000, 5.
2. B. Baker, *Environ. Action*, 1995, *27*(1), 9.
3. (a) G.T. Gardner and P.C. Stern, *Environmental Problems and Human Behavior*, Allyn & Bacon, Boston, 1996; (b) C.G. Herndl and S.C. Brown, eds, *Green Culture— Environmental Rhetoric in Contemporary America*, University of Wisconsin, Madison, WI, 1996; (c) K.I. Noorman and T.S. Uiterkamp, eds, *Green Households? Domestic Consumers, the Environment and Sustainability*, Earthscan, London, 1997; (d) G Gardner and P. Sampat, *Mind Over Matter: Recasting the Role of Materials in Our Lives*, World Watch Institute paper 144, Washington, DC, 1998; (e) R. Rosenblatt, ed., *Consuming Desires—Consumption, Culture and the Pursuit of Happiness*, Island Press/Shearwater Books, Washington, DC, 1999.
4. J. John, *Simple Things You Can Do to Save the Earth*. Earthworks Press, Berkeley, CA, 1989 (some typical actions that a consumer can take).
5. J. Makower, *Wilmington Delaware News J.*, 1995, D3.
6. (a) D.A. Coleman, *Ecopolitics: Building a Green Society*, Rutgers University Press, New Brunswick, NJ, 1994; (b) S.P. Hays, *A History of Environmental Politics Since 1945*, University of Pittsburgh Press, Pittsburgh, 1999.
7. Anon., *Environ. Sci. Technol.*, 1996, *30*, 160A.
8. R.M. White, *Technol. Rev.*, 1996, *99*(2), 56.
9. (a) J. Davis, *Amicus J. (Natural Resources Defense Council)*, 1995, *17*(2), 18; (b) *Amicus J. (Natural Resources Defense Council)*, 1995, *17*(2), 3; (c) K. Durbin, *Amicus J. (Natural Resources Defense Council)*, 1995, *17*(3), 29; (d) D.N. Foley and K.D. Lassila, *Amicus J. (Natural Resources Defense Council)*, 1995, *17*(3); (e) Anon., *Amicus J. (Natural Resources Defense Council)*, 1995, *17*(3), 3; (f) D. Meadows, *Amicus J. (Natural Resources Defense Council)*, 1996, *18*(1), 12; (g) Anon., *Amicus J. (Natural Resources Defense Council)*, 1996, *18*(1), 55; (h) D.J. Hanson, *Chem. Eng. News*, July 24, 1995, 32; (i) P. Goldman, *Environment*, 1996, *38*(3), 41; (j) B. Hileman, *Chem. Eng. News*, Mar. 11, 1996, 8.
10. A. Gore. *Earth in the Balance—Ecology and the Human Spirit*. Houghton Mifflin, Boston, 1992.
11. (a) M. Jacobs, *Chem. Eng. News*, Oct. 9, 1995, 5; (b) B. Hileman, *Chem. Eng. News*, June 19, 1995, 21; (c) R.M. Baum, *Chem. Eng. News*, Aug. 31, *20-09*, 3
12. D.N. Foley and K.D. Sassila, *Environ. Sci. Technol.*, 1996, *30*, 13.
13. G. Peaff, *Chem. Eng. News*, July 7, 1997, 27.
14. B. Hileman, *Chem. Eng. News*, Feb. 22, 1999, 23.
15. W. Rudig, *Environment*, 2002, *44*(3), 20.
16. Anon., *Chem. Ind. (Lond.)*, 1996 397; 2000, 49; 1999, 123.
17. B. Hileman, *Chem. Eng. News*, June 14, 1999, 21.
18. (a) J. Werksmann, *Greeting International Institutions*. Earthscan, London, 1995; (b) O.R. Young, *Environment*, 1999, *41*(8), 20.

19. D. Hinrichsen, *Amicus J. (Natural Resources Defense Council)*, 1996, *18*(3), 35.

20. W.C. Clark, *Environment*, 1997, *39*(7), 1.

21. World Commission on Environment and Development, *Our Common Future*. Oxford University Press, Oxford, 1987.

22. (a) J.B. Callicott and F.J.R. da Rocha, eds, *Earth Summit Ethics: Toward a Reconstructive Postmodern Philosophy of Environmental Education*, State University of New York, Albany, NY, 1996; (b) T. Heyd, *Environ. Ethics*, 1997, *19*, 434; (c) M. McCoy and P. McCully, In: I. Tellam and P. Chatterjee, eds, *The Road from Rio: An NGO Action Guide to Environment and Development*, International Books, Utrecht, 1993.

23. B. Hileman, *Chem. Eng. News*, Sept. 2, 2002, 11.

24. (a) E.B. Weiss and H.K. Jacobson, *Environment*, 1999, *41*(6), 16; (b) E.B. Weiss and H.K. Jackson, eds, *Engaging Counties: Strengthening Compliance with International Environmental Accords*, MIT Press, Cambridge, MA, 1998.

25. B. Hileman, *Chem. Eng. News*, Sept. 21, 1998, 41.

26. (a) T.G. Weiss and L. Gordenker, eds, *NGOs, the U.N. and Global Governance*, L. Rienner, Boulder, CO, 1996; (b) T. Prince and M. Finger, *Environmental NGOs in World Politics: Linking the Local and the Global*, Routledge, London, 1994; (c) P. Willetts, ed., *Conscience of the World: The Influence of NGOs in the UN System*, Hurst, London, 1996; (d) A. Fowler, *Strikling a Balance—A Guide to Enhancing the Effectiveness of Non-Governmental Organizations in International Development*, Earthscan, London, 1997.

27. (a) R.M. Baum, *Chem. Eng. News*, Oct. 30, 1995, 5; (b) J. Darabaris, *Corporate Environmental Management*, CRC Press, Boca Raton, Florida, 2008.

28. (a) D. Rejeski, *Technol. Rev.*, 1997, *100*(1), 56; (b) R. Breslow, *Chem. Eng. News*, Aug. 26, 1996, 72; (c) B.R. Allenby and D.J. Richards, *The Greening of Industrial Ecosystems*, National Academy Press, Washington, DC, 1994, (d) F. Cairncross, *Green Inc.—A Guide a Business and the Environment*, Earthscan, London, 1995; (e) J.D.Graham and J.K. Hartwell, *The Greening of Industry—A Risk Management Approach*, Harvard University Press, Cambridge, MA, 1997; (f) P. Groenewegen, K. Fischer, E.G. Jenkins, and J. Schot, eds, *The Greening of Industry Resource Guide and Bibliography*, Island Press, Washington, DC, 1995; (g) R.A. David. *The Greening of Business*, Gower, Brookfield, VT, 1991. (h) K.M. Reese, *Chem. Eng. News*, Oct. 2, 2000, 200. (i) P.G. Derr and E.M. McNamara, *Case Studies in Environmental Ethics*, Rowman and Littlefield Publishers, Lanham, MD, 2003.

29. R.E. Chandler, In: K. Martin and T.W. Bastock, *Waste Minimisation: A Chemist's Approach*, Royal Society of Chemistry, Cambridge, 1994, 62–74.

30. M.D. Rogers. *Business and the Environment*, Macmillan, Basingstoke, England, 1995, 3.

31. V.N. Bhat, *The Green Corporation—The Next Competitive Advantage*. Quorum Books, Westport, CO, 1996.

32. (a) K. Fischer and J. Schot, eds, *Environmental Strategies for Industry*, Island Press, Washington, DC, 1993; (b) T. Saunders and L. McGovern, *The Bottom Line of Green Is Black*, Harper, San Francisco, 1994.

33. (a) D.J. Schell, *A Green Plan for Industry: 16 Steps to Environmental Excellence*, Government Institutes, Rockville, MD, 1998; (b) T. Chapman, *Chem. Ind. (Lond.)*, 1998, 834; (c) L.D. DeSimone and F. Popoff, *Eco-Efficiency: The Business Link to Sustainable Development*, MIT Press, Cambridge, MA, 1997; (d) N.J. Roome, *Sustainability Strategies for Industry—the Future of Corporate Practice*, Island Press, Washington, DC, 1998; (e) K. Fischer and J. Schot, *Environmental Strategies for Industry—International Perspectives on Research Needs and Policy Implications*, Island Press, Washington, DC, 1993; (f) R. Howes, J. Skea, and B. Whelan, *Clean and Competitive? Motivating Environmental Performance in Industry*, Earthscan, London,

1998; (g) P.M. Morse, *Chem. Eng. News*, Aug. 3, 1998, 13; (h) D. Gaskell, *Chem. Br.*, 200, *36*(6), 36; (i) A.L. White, *Environment*, 1999, *41*(8), 30; (i) A.J.Hoffmann, *Environment*, 2000, *42*(5), 22; (k) M. Arnold, *Chem. Ind. (Lond.)*, 1999, 921; (l) N.J. Roome, *Sustainability Strategies—The Future of Corporate Practice*, Island Press, Washington, DC, 1998; (m) C. Laszlo, *The Sustainable Company: How to Create Lasting Value Through Social and Environmental Performance*, Island Press, Washington, DC, 2003.

34. P. Hawken, *The Ecology of Commerce—a Declaration of Sustainability*, Harper, San Francisco, 1993, 31.

35. (a) D. Hanson, *Chem. Eng. News*, May 8, 1995, 24; (b) *Chem. Eng. News*, May 15, 1995, 17; (c) W. Storck, *Chem. Eng. News*, Oct. 5, 1998, 26.

36. Anon., *Chem. Eng. News*, May 15, 1995, 22.

37. D. Hanson, *Chem. Eng. News*, Apr. 7, 1997, 12.

38. (a) D. Hanson, *Chem. Eng. News*, Aug. 7, 1995, 8; (b) T. Agres, *R&D (Cahners)*, Dec. 1996, *38*, 29.

39. (a) J.L. Grisham, *Chem. Eng. News*, June 7, 1999, 21; (b) G. Parkinson, *Chem. Eng*, 2001, *108*(10), 25.

40. D. Hanson, *Chem. Eng. News*, Apr. 26, 1999, 10.

41. (a) Anon., *Chem. Eng. News*, Feb. 24, 1997, 25; (b) Anon., *Chem. Eng. News*, Sept. 14, 1998, 21.

42. Anon., *Environ. Sci. Technol.*, 1997, *31*, 124A.

43. (a) B. Smart, ed., *Beyond Compliance: A New Industry View of the Environment*, World Resources Institute, Washington, DC, 1992; (b) P. Layman, *Chem. Eng. News*, Oct. 18, 1999, 33; (c) C. Holliday, S. Schmidheiny, and P. Watts, *Walking the Talk*, Greenleaf Publishing, Sheffield, UK, 2002.

44. (a) Anon., *Chem. Ind. (Lond.)*, 1995, *164*, 906; (b) B. Dale, *Chem. Ind. (Lond.)*, 1994, 976.

45. L.B. Lave and H.S. Matthews, *Technol. Rev.*, 1996, *99*(8), 68.

46. (a) D.C. Korten, *When Corporations Rule the World*, Kumarian Press, West Hartford, CO, 1995, 146–156; (b) D. Helvarg, *Amicus J. (Natural Resources Defense Council)*, 1996, *18*(2), 13; (c) D. Edward, *Ecologist*, 1999, *18*(2), 13; (c) D. Edward, *Ecologist*, 1999, *29*(3), 172; (d) F. Simon and M. Woodell, In: D.J. Richards, ed., *The Industrial Green Game—Implications for Environmental Design and Management*, Royal Society of Chemistry, Cambridge, UK, 1997, 212–224.

47. Anon., *Chem. Ind. (Lond.)*, 1998, 473.

48. (a) D. Fagin and M. Lavelle, *Toxic Deception: How the Chemical Industry Manipulates Science, Bends the Law and Endangers Your Health*, Common Courage Press, Monroe, ME, 1999; (b) G. Markowitz and D. Rosner, *Deceit and Denial—The Deadly Politics of Industrial Pollution*, University of California Press, Berkeley, CA and Milbank Memorial Fund, New York, 2002; (c) D. Michaels, *Doubt Is Their Product—How Industry's Assault on Science Threatens Your Health*, Oxford University Press, New York, 2008; (d) W. Freudenberg, R. Gramling, and D.Davidson, *World Watch*, 2008, *21*(3), 7.

49. M. Heylin, *Chem. Eng. News*, June 2, 1997, 12; May 31, 1999, 15.

50. (a) G. Peaff, *Chem. Eng. News*, Nov. 25, 1996, 23; (b) J.L. Wilson, *Chem. Eng. News*, June 3, 1996, 5; (c) A.R. Hirsig, *Chem. Eng. News*, Sept. 8, 1997, 5.

51. (a) F.L. Webber, *Chem. Eng. News*, Nov. 9, 1998, 5; (b) J. Johnson, *Chem. Eng. News*, Nov. 2, 1998, 19; Mar. 8, 1999, 9.

52. R.S. Rogers, *Chem. Eng. News*, Apr. 12, 1999, 30.

53. E.H. Fording Jr., *Chem. Eng. News*, May 4, 1998, 5.

54. (a) M. McCoy, *Chem. Eng. News*, Aug. 2, 1999, 16; (b) A.M. Thayer, *Chem. Eng. News*, June 8, 1998, 17.

55. D.J. Hanson, *Chem. Eng. News*, July 15, 1996, 29.
56. Anon., *Chem. Eng. News*, Aug. 18, 1997, 33.
57. Anon., *Chem. Eng. News*, May 19, 1997, 29.
58. G. Parkinson, *Chem. Eng.*, 1996, *103*(11), 27.
59. (a) J. Johnson, *Chem. Eng. News*, July 28, 1997, 30; June 30, 1997, 26; (b) Anon., *Chem. Eng. News*, Dec. 9, 1996, 23; June 30, 1997, 25; Dec. 22, 1997, 19; (c) Anon., *Environ. Sci. Technol.*, 1998, *32*, 129A.
60. Anon., *Chem. Eng. News*, Mar. 10, 1997, 33.
61. A. Thayer, *Chem. Eng. News*, Apr. 13, 1998, 32.
62. (a) M. Jacobs, *Chem. Eng. News*, Nov. 24, 1997, 5; (b) Anon., *Chem. Ind. (Lond.)*, 1996, 961.
63. M. McCoy, *Chem. Eng. News*, Sept. 26, 2005, 10; June 26, 2006, 27.
64. P. Barry, *AARP Bull.*, Washington, DC, June 2002, 8.
65. (a) E. Kirchner, *Chem. Eng. News*, Apr. 22, 1996, 23; (b) L.R. Ember, *Chem. Eng. News*, May 29, 1995, 10; (c) M. Heylin, *Chem. Eng. News*, May 29, 1995, 5; (d) A. Shanley, G. Ondrey, and J. Chowdhury, *Chem. Eng.*, 1997, *104*(3), 39; (e) M.S. Reisch, *Chem. Eng. News*, Jan. 12, 1998, 104; May 11, 1998, 13; Oct. 26, 1998, 15; May 24, 1999, 18; Sept. 4, 2000, 21; (f) D. Hunter, ed., *Chem. Week*, July 5/12, 2000, 38–85; (g) R. Stevenson, *Chem. Br.*, 1999, *35*(5), 27; (h) F.M. Lynn, G. Busenberg, N. Cohen, and C. Chess, *Environ. Sci. Technol.*, 2000, *34*, 1881.
66. J.-F. Tremblay, *Chem. Eng. News*, Oct. 21, 1996, 21; June 23, 1997, 20.
67. Anon., *Chem. Ind. (Lond.)*, 1998, 872.
68. (a) M.S. Reisch, *Chem. Eng. News*, Apr. 14, 1997, 21; (b) M.S. Reisch, *Chem. Eng. News*, May 26, 2003, 15; (c) P. Short, *Chem. Eng. News*, May 26, 2003, 19.
69. K.S. Betts, *Environ. Sci. Technol.*, 1998, *32*, 303A.
70. (a) P. Layman, *Chem. Eng. News*, Dec. 22, 1997, 9; Jan. 5, 1998, 17; (b) Anon., *Chem. Ind. (Lond.)*, 1998, 3; (c) Anon., *Chem. Eng. News*, Mar. 1, 1999, 20.
71. (a) S.J. Bennett, R. Freierman, and S. George, *Corporate Realities and Environmental Truths: Strategies for Leading Your Business in the Environmental Era*, Wiley, New York, 1993, 6, 19, 22, 24; (b) J. Nash and J. Ehrenfeld, *Environment*, 1996, *38*(10), 16.
72. G. Hess, *Chem. Eng. News*, Jan. 15, 2009, 19.
73. C. Hogue, M.P. Walls, and J. Tickner, *Chem. Eng. News*, Jan. 8, 2007, 34.
74. (a) C. Hogue, *Chem. Eng. News*, Sept. 22, 2008, 12; Oct. 6, 2008, 42; (b) M. Murphy, *Chem. Ind. (Lond.)*, May 5, 2008, 5.
75. C. Hogue, *Chem. Eng. News*, May 12, 2008, 9; (b) Anon., *Chem. Eng. News*, July 28, 2008, 41; (c) C. Hogue, *Chem. Eng. News*, June 1, 2009, 4.
76. (a) J. Johnson, *Chem. Eng. News*, Apr. 7, 2008, 10; (b) C. Hogue and J. Johnson, *Chem. Eng. News*, Apr. 14, 2008, 35.
77. Anon., *Chem. Eng. News*, Oct. 17, 2008, 27.
78. (a) Anon., *Amicus J. (Natural Resources Defense Council)*, 1990, *12*(2), 3; (b) V.N.Bhat, *The Green Corporation—The Next Competitive Advantage*, Quorum Books, Westport, CO, 1996, 163; (c) http://www.ceres.org; (d) K. Betts, *Environ. Sci. Technol.*, 1999, *33*, 189A.
79. T.F. Walton, *Environ. Prog.*, 1996, *15*(1), 1.
80. S.A. Fenn, *Technol. Rev.*, 1995, *98*(5), 62.
81. (a) D. Green, *ISO 9000 Quality Systems Auditing*, Gower, Aldershot, UK, 1997; (b) W.A. Golomski, A.J.M. Pallett, J.G. Surak, and K.E. Simpson, *Food Technol.*, 1994, *48*(12), 57, 60, 63; (c) A. Badiru, *Industry's Guide to ISO 9000*, Wiley, New York, 1995.
82. (a) D. Hunt and C. Johnson, *Environmental Management Systems—Principles and Practice*, McGraw-Hill, New York, 1995; (b) J. Cascio, G. Woodside, and P. Mitchell, *ISO 14000 Guide: The New International Management Standards*, McGraw-Hill,

New York, 1996; (c) J.M. Diller, *Chem. Eng. Prog*, 1997, *93*(11), 36; (d) R. Begley, *Environ. Sci. Technol.*, 1996, *30*(7), 298A; 1997, *31*(8), 364A; (e) G.S. Samdani, S. Moore, and G. Ondrey, *Chem. Eng.*, 1995, *102*(6), 41; (f) A.M. Thayer, *Chem. Eng. News*, Apr. 1, 1996, 11; Sept. 30, 1996, 27; (g) B. Rothery, *ISO 14000 and ISO 9000*, Gower Publishing, Aldershot, UK, 1995; (h) W.M. von Zharen, *ISO 14000: Understanding the Environmental Standards*, Government Institutes, Rockville, MD, 1996; (i) J. Cascio and J.S. Shideler, *Chemtech*, 1998, *28*(5), 49; (j) A. Schoffman and A. Tordini, *ISO 14001: A Practical Approach*, American Chemical Society, Washington, DC, 2000; (k) P.J. Knox, *Chem. Process (Chicago)*, 1999, *62*(2), 26; (l) R. Krut and H. Gleckman, *ISO 14001—The Missed Opportunities*, Earthscan, London, 1998; (m) N.I. McClelland and B. St. John, *Environ. Prog.*, 1999, *18*(1), S3.

83. Anon., *Chem. Eng. News*, Jan. 29, 1996, 15.
84. J.W. Houck and O.F. Williams, eds, *Is the Good Corporation Dead?* Rowman & Littlefield, London, 1996.
85. A. Thayer, *Chem. Eng. News*, May 4, 1998, 31.
86. (a) Anon., *Chem. Eng. News*, Aug. 30, 1999, 39; (b) Anon., *Industrial Environmental Performance Metrics: Challenges and Opportunities*, National Academy of Engineering, Washington, DC, 1999; http://national-academies.org.
87. V.N. Bhat, *The Green Corporation—The Next Competitive Advantage*. Quorum Books, Westport, CT, 1996, 165.
88. M. Freemantle, *Chem. Eng. News*, May 20, 1996, 30.
89. (a) Anon., *Chem. Ind. (Lond.)*, 1994, 928; (b) Council on Economic Priorities, New York, 1995.
90. S.J. Bennett, R. Freierman and S. George, *Corporate Realities and Environmental Truths:Strategies for Leading Your Business in the Environmental Era*, Wiley, New York, 1993, 156.
91. F. Krupp, *EDF Lett. (Environmental Defense Fund)*, 1996, *27*(4), 4.
92. Chevron, *Measuring Progress—A Report on Chevron's Environmental Performance*, San Francisco, CA, Oct. 1994.
93. Anon., *Chem. Eng.*, 1999, *106*(4), 64.
94. G. Ondrey, *Chem. Eng.*, 2000, *107*(11), 27.
95. R. Rivera, *Amicus J. (Natural Resources Defense Council)*, 2001, *2*(4), 10.
96. Chevron, 1997, Annual Report, San Francisco, CA.
97. K. Koenig, *World Watch*, 2004, *17*(1), 10.
98. J. Diamond, *Conserv. Pract.*, 2005, *6*(4), 12.
99. Chevron, *Quarterly Reports to Shareholders for the Third Quarters of 1996 and 1997*, San Francisco, CA.
100. (a) Earth Day 2000, *The Don't Be Fooled Report—the Top Ten Greenwashers of 1994*, San Francisco, CA, 1995; (b) D. Helvarg, *Amicus J. (Natural Resources Defense Council)*, 1996, *18*(2), 16.
101. Chevron, Report to shareholders for the second quarter of 1995.
102. (a) Chevron, *1995 Meeting Report*, San Francisco, CA, June 10, 1995; (b) Chevron, *Report to Shareholders*, San Francisco, CA, first quarter 1997; (c) Chevron, *Proxy Statement for Its Annual Shareholders Meeting*, San Francisco, CA, Apr. 1999.
103. (a) J.H. Krieger, *Chem. Eng. News*, July 8, 1996, 13; (b) Green Chemistry and Engineering Conference, Washington, DC, June 23–25, 1997; (c) P.V. Tebo, *Chemtech*, 1998, *28*(3), 8.
104. B. Hileman, *Chem. Eng. News*, July 30, 2001, 13.
105. B. Hileman, *Chem. Eng. News*, Nov. 23, 1998, 10.
106. R.E. Chandler, In: K. Martin, T.W. Bastock, eds, *Waste Minimisation—A Chemist's Approach*, Royal Society of Chemistry, Cambridge, 1994, 71.

107. (a) Anon., *Chem. Eng. News*, Aug. 29, 1994, 21; (b) G. Samdani, *Chem. Eng.*, 1995, *102*(5), 19; (c) Anon., *Chem. Ind. (Lond.)*, 1994, 668; (d) Anon., *Chem. Eng. News*, Jan. 9, 1995, 9; (e) Anon., *Chem. Eng. News*, Apr. 8, 1996, 36; (f) Anon., *Chem. Eng. News*, July 22, 1996, 36; (g) E.R. Beaver, *Environ. Prog.*, 1997, *16*(3), F3.

108. The Presidential Green Chemistry Challenge Awards Program—Summary of 1996 Award Entries and Recipients. EPA 744-K-96-001, U. S. Environmental Protection Agency, Washington, DC, July 1996, 2.

109. D. Fagin and M. Lavelle, *Toxic Deception: How the Chemical Industry Manipulates Science, Bends the Law and Endangers Your Health*, Carol Publishing, Secaucus, NJ, 1996.

110. (a) A. Thayer, *Chem. Eng. News*, Dec. 22, 1997, 18; (b) M.S. Reisch, *Chem. Eng. News*, Jan. 12, 1998, 106.

7 Solutions to In-Chapter Problems

SOLUTIONS TO CHAPTER 1: TOXICITY, ACCIDENTS, AND CHEMICAL WASTE

Solution 1.1 Pollution Prevention at an Isocyanates Plant

There are situations where this type of plan has and has not been adopted. For example, the plan was not implemented when a company's business center thought it could make more money by investing the money elsewhere. This was at a Dow plant in Texas, and involved the Natural Resources Defense Council. Later, a nongovernmental organization did the same audit for the Dow plant in Midland, Michigan, where the annual savings would be US$4 million. This one was implemented, since the company was able to reduce capital expenses by using second-hand equipment.[1]

Solution 1.2 Curious Polar Bears

A thin layer of a pesticide on a leaf can sublime into the air and then travel long distances until rain or cold weather brings it to earth. Analysis of the blood of the polar bears by techniques such as capillary gas chromatography and mass spectrometry has found highly chlorinated insecticides. Highly brominated flame retardants have also been known to travel long distances, and highly chlorinated insecticides are persistent in the environment. There is nothing that can be done for these particular polar bears. The Stockholm Convention on persistent organic pollutants can prevent further accumulation in distant places. Countries that ratify the treaty agree to outlaw the use of nine persistent organic pollutants. Dichlorodiphenyltrichloroethane (DDT) was exempted since it is still used to fight malaria in some countries. The treaty went into force in 2004. As of 2013, 178 countries have ratified the treaty, although the United States is not one of them. There are similar problems in the tropics. The black-footed albatrosses of Midway Island in the Pacific Ocean are having reproductive problems.[2]

Solution 1.3 After an Industrial Accident

This happened a few years ago in Delaware City, Delaware.

- Devise a new process that does not use chlorine.
- Devise a process where chlorine is generated in situ as needed, so that only a small amount is present at any time.
- Use a double-walled tank.
- Pressure-test any new tank with nitrogen or water to be sure it can handle the pressure to be used. Accidents have happened with natural gas testing, such as one in Connecticut that led to a fire and explosion.

- Inspect the tank for poor welds or other defects by ultrasound, and repeat the inspection every few months. Pipelines have failed due to lack of inspections in Alaska and other places.
- Put a maximum pressure control on the tank, in case the accident was caused by over-pressuring.
- Provide better training for the operators.
- Be sure that the tank is made of a material that will not corrode with chlorine.
- Can the chlorinated product be replaced by one not containing chlorine?

Solution 1.4 A Mighty Safety Dilemma

There are many correct answers to this problem. It is significant that students select the proactive ones, and none that penalize anyone.

Solution 1.5 To Burn or Not to Burn?

- Ask the company why it cannot use a different site located further from the houses.
- Ask the company why the stack is not taller.
- Require trial burns to check for Hg, fine particulates, and chlorodioxins.
- Ask the company if biohazardous waste is to be included. If so, will all dangerous microbes be destroyed? No radioactive material can be allowed in the waste.
- Ask the company why they have to use chlorinated solvents, and if they can be replaced.

Solution 1.6 "Delacid": A Versatile Catalyst

- Start with nondestructive approaches, and then move to ones that require very small samples, such as mass spectrometry and thermal methods.
- X-ray diffraction, nuclear magnetic resonance, infrared spectroscopy, surface area analysis, porosity, and pyrolysis gas chromatography can be used.
- The insolubility claims can easily be checked.
- The acidity can be measured by adsorption of amines.
- The catalyst effectiveness can be determined by trying it in typical test reactions.

Students tend to skip elemental analysis of the catalyst, not realizing that the sample might be a mixture.[3–5]

Solution 1.7 The Flask Broke

- Ask how dry the solvent has to be for its intended use. Many reactions can tolerate a trace of water (cationic polymerizations require it).
- The solvent could be stored overnight over molecular sieves and transferred with a cannula to the nitrogen-filled flask.
- Use a column of alumina with collection under nitrogen gas.

- Sparging with nitrogen gas for twenty minutes will get the water content very low.
- Purchase dry solvent from Sigma-Aldrich or another company.

Students may have some other approaches such as using a heavier glass flask, or using a stainless steel flask.

Solution 1.8 The Tremendous Problem of Climate Change

- This may require energy from solar and wind power.
- Make existing polymers starting with ethylene from ethanol that is derived from sugar cane in Brazil. Dow is preparing to do this.
- Find better and cheaper ways to prepare derivatives from cellulose, hemi-cellulose, lignin, and chitin. This might be done in ionic liquids or in an extruder.
- Convert biomass to synthesis gas, which can be used to make monomers.
- Make diols and dicarboxylic acids starting with hydroformylation of oleic acid, followed by reduction to form the diols. The reaction of the two produces a polyester.
- Convert glucose to sorbitol and then to isosorbide. This is a diol that can be reacted with a dicarboxylic acid, such as adipic acid, which is obtainable from glucose by using a chemoenzymatic method. The rings will raise the glass transition temperature and the melting point.
- Use succinic acid from fermentation. It can be converted to 1,4-butanediol and 1,4-diaminobutane. These can be reacted with more succinic acid to form polyesters and polyamides.

Solution 1.9 The Chemist Talks to the Chemical Engineer

- The price of the starting material is way too high. The chemist should hunt for a starting material that can be obtained in bulk from a major manufacturer.
- The potassium dichromate will produce large amounts of toxic residues. He should see whether a newer oxidation with air, ozone, or hydrogen peroxide could be used.
- Diazomethane is too hazardous a reagent to use on a large scale. Is an alternative possible?
- The chemist should see if a catalytic reduction with hydrogen gas can be used in place of the expensive lithium aluminum hydride, which has to be used on a per mole basis. Diethyl ether requires all explosion-proof switches in the plant.
- Protection and deprotection steps are to be avoided if at all possible.
- Dioxane and 2-methoxyethanol are teratogenic solvents and should not be used.
- Mercuric acetate is too toxic to use.
- A 24-hour reaction at -78°C is too long, and involves costly refrigeration.

- A pressure of 7 kbar requires special expensive equipment and remote operation.
- The distillation is at a pressure that is not available on a plant scale, and the 100 plate columns would never be used in a commercial facility. Another purification method is needed.

The whole route needs to be rethought by the chemist and chemical engineer before any further laboratory work is undertaken.

Solution 1.10 Inherently Safer Chemistry

Industry is probably afraid that inherently safer chemistry will cost them more money. It could if a whole new production plant has to be built. Plants that can implement inherently safer chemistry with minor changes may have done so already. This has been achieved in eliminating the shipment of chlorine for use in the treatment of water and wastewater. Sodium hypochlorite can be shipped instead, or the chlorine can be generated onsite in a turnkey plant. It might be a major change if, for example, a plant had to shift from electroplating to chemical vapor deposition, or if it made its ethanol by fermentation instead of adding water to ethylene.

Companies have added more guards, fences, and swipe cards. They seem to forget that the terrorists of 9/11/01 came by air. A disgruntled hunter used a high power rifle to shoot a hole in the Alaska Pipeline. The Massachusetts law requires companies to think through their processes from start to finish, but does not require them to actually implement the new methods. However, many of them have done so and have saved millions of dollars. New Jersey has a similar law.

Solution 1.11 A Strange Malady

Rachel Carson described this problem in her book *Silent Spring* (Houghton Mifflin, 1962) and testified before Congress about it. She was subject to strong criticism by industry and even by *Chem. Eng. News*, who questioned her data and her method of getting it. She has since been vindicated because she collected her data carefully and was correct. There is now a national wildlife refuge in Maine bearing her name.

- The problem was due to eggshell thinning, which caused the eggs to break. This was due to DDT and other highly chlorinated pesticides.
- Eggs from less contaminated areas hatched normally in the problem areas. Captive breeding had to be used with the peregrine falcons. The pesticides were banned in the United States. Today, the species have recovered.[6]

Solution 1.12 The Mysterious Case of the Disappearing Filter Paper

- Replace the polluting xanthate process for making rayon that uses carbon disulfide as both reagent and solvent.
- Run reactions on cellulose to make derivatives that can better compete with petroleum-based polymers for a sustainable future.

- Find a cheaper, nontoxic ionic liquid for commercial use. This might be a deep eutectic solvent based on choline and an amino acid.
- Patent and license it to a chemical company.

This discovery has led to a new method of making rayon. It also offers the possibility of cheaper cellulose derivatives to compete better with polymers made from petroleum and natural gas. BASF has licensed the process and has made cellulose acetate in ionic solvents. Many other reactions of cellulose in ionic liquids should be possible, such as making other esters, ethers, and so on. It might also be used for degradation of cellulose to sugars for use in making biofuels.[7]

SOLUTIONS TO CHAPTER 2: THE CHEMISTRY OF LONGER WEAR

Solution 2.1 I Lost My Pants

- Analyze the rubber to see if there any stabilizers present. If so, see if they are still there after washing and drying several times.
- Substitute rubbers such as butyl rubber or a polyurethane rubber with one that has many fewer or no double bonds.
- Try multiple grippers, multiple buttons, velcro, drawstring, and suspenders.
- Wear a nightgown instead of pajamas!

Solution 2.2 My Pantyhose Ran Faster than the Runners in the Boston Marathon

- Make them heavier with a larger denier thread.
- Use a fiber stronger than nylon, for example, Kevlar, Nomex, or ultrahigh molecular weight polyethylene.
- Wear no pantyhose or stockings, or wear pants.
- Use a woven fabric instead of nylon, which is knitted.
- Use the techniques for holding nonwoven fabrics together, such as ultrasound, heat on bicomponent fibers, mixtures of lower melting and higher melting fibers plus heat, and adhesives that migrate to joints.
- Use dihydroxyacetone to tan the legs, as dermatologists say that there is no such thing as a healthy tan.
- Place the pantyhose in a freezer before wearing. Freezing strengthens nylon fibers against runs and pulls.
- Keep fingernails well-trimmed!

Solution 2.3 The Clothes Horse

The statistics come from the book *Overdressed: The Shockingly High Cost of Cheap Fashion* by Elizabeth Cline (Portfolio Trade, 2012). The author has 364 pieces of clothing in her closet. This was not the case before 1920, according to *Waste and Want: A Social History of Trash, a book by Susan Strasser (Holt Paperbacks, 2000)*. Before 1920, most clothing was mended, altered for other family members, or recycled in the home as braided throw rugs or quilts. Consignment shops, yard sales, and clothing exchanges are not enough to solve the problem.

- Consider reversible clothes where both sides can be used for variety.
- How might the length be varied from miniskirt to full-length gown without any cutting required?
- Do we really need suit jackets and ties in summer? Must we be as formal as we often are?
- Women have twice as many foot problems as men, probably because of pointed high heel shoes. Are these necessary?
- Polyethylene terephthalate (PET) is the fiber used the most; it is manufactured from petroleum. Should we find ways to make it from renewable resources? Considering the environmental problems with cotton, perhaps clothes should be made of linen or hemp.

Solution 2.4 Tires on and in Rubber

- It may slow construction. Various particle and surface treatments may give different results. It can only be applied in warm weather when the temperature is no lower than 55°F. About 18%–20% ground rubber is used.
- This avoids the problems of insects breeding in the water of scrap tires and the fires that have happened in piles of old tires.[8]

Solution 2.5 An Opinion Poll on Clothing and Other Items

1. a. Powder coatings work fine on appliances as there is no overspray. All of the powder can be used. There is no solvent, as there is in the usual car finish. A "flat" finish on a car can make driving easier on a sunny day since it reduces the glare from the hood.
   b. The chromium-plated bumper is a thing of the past. It was too expensive and customers were unwilling to pay a premium for it.
   c. People still seem to want a shiny floor. Such floors are no longer slippery since they contain an anti-slip agent (possibly silica).
   d. Fingernails usually look fine without any paint and require no volatile solvent.
   e. Polishes for shoes contain toxic nitrobenzene solvent. Can it be replaced by a harmless solvent? Shoes do not have to be shiny to look good and there will be no air pollution.
   f. Lighting fixtures need a coating of lacquer to keep them shiny.
   g. Earrings do not have to be shiny, but they usually are.
   h. Shiny pants should be admired since it indicates an individual who cares enough about the environment to wear his clothes as long as he can. It could require some mending.
2. The usual crease-proofing agent that cross-links the cellulose can release the carcinogen formaldehyde. Butane-1,2,3,4-tetracarboxylic acid releases only water as it cures, but it costs more.
3. Fashions change. The rumpled look could become popular and eliminate the need for crease-proofing. Some hair styles are very rumpled.
4. There is no need for suits—just a requirement that people dress neatly. Suits use more material and energy to make them, and require more money

to buy them. Since they can be hot, the amount of laundry goes up. The amount of air conditioning can go up to keep the suited people comfortable.

5. Brick does not have to be painted, but it has become expensive to lay all the bricks by hand. Natural wood protectors only last one or two seasons. Aluminum can make unwanted noises on expansion and contraction. Poly(vinyl chloride) is made from a carcinogenic monomer.

6. The dry cleaning solvent perchloroethylene is being phased out since it may be harmful to people. It is easy to buy clothing that does not need to be dry-cleaned.

7. Do not collect the clippings. Leave them on the ground or put them into your compost pile, or have less lawn.

8. All hair sprays contain an organic solvent.

9. An outdoor clothesline works well and reduces the need for energy in a dryer. It should be used in the suburbs, but some deed restrictions prohibit it. Using cold water detergent to save energy does not do much good if the clothes have to be dried in an indoor dryer.

10. Use muscle power and keep slim and fit. No fossil fuels are needed. A home owner with a small yard does not need all these gadgets. You may be able to drop your membership at the gym and save money. You could also get more exercise by using a reel mower that you push yourself.

SOLUTIONS TO CHAPTER 3: THE CHEMISTRY OF WASTE MANAGEMENT AND RECYCLING

Solution 3.1 Cost-Effective Handling of Sewage Waste

- Talk to the environmentalists and find out why they are complaining.
- Ask them how new methods, such as secondary and tertiary treatment, should be funded. See if they would put up the money.
- See what independent estimates of the cost would be, in concert with hiring a consultant.

This happened in the city of Philadelphia. The federal government stated that they would put up 80% of the required funding. A major problem was that the chemical oxygen demand provided an anaerobic zone that fish could not cross. The impasse was ended when the states of Delaware and New Jersey joined with an environmental organization and brought a lawsuit against the city. A judge ordered the city to clean up.

Solution 3.2 The Computer Age

- Develop a cell phone that will last for ten years using no toxic materials.
- Make a computer without using toxic materials.
- Develop a device where the housing can be reused by just adding a new chip with the latest technology.
- Require manufacturers to take back what they sell at the end of its useful life.
- Include the cost of disposal in the initial price.
- Develop a system of recycling centers for electronic equipment.
- Prohibit export of electronic waste to developing nations.

- Devise simpler recycling methods.
- Donate old computers to a school or charity.[9]

Solution 3.3 A Printing Challenge

- Solvents can be burned with a gas burner, but this would not be efficient with the large volumes of air needed (neither would a cold trap). Zeolites could adsorb the solvents and release them on heating to be reused.
- Printing can be done with light, or curing by an electron beam. The processes require special monomers containing acrylates or methacrylates and are used widely.

Solution 3.4 It Is Turkey Time

- Convert the waste to biogas, and use it to heat and cool the turkey house and your house.
- Burn the waste for energy that can be used to make electricity.
- Ship the waste back to where the corn and soybeans are grown to be used as fertilizer. Heat and sterilize the waste and sell to home gardeners for fertilizer. Milwaukee sells sewage sludge as Milorganite, and Vermont sells cow feces as Moo Doo.
- Convince the state to set up an exchange program to move the waste from where there is too much to where there is not enough.
- Use a state subsidy to pyrolyze the waste and sell the residue as fertilizer. Use the waste gases to help power the plant.

Solution 3.5 Toys and Games from Trash

Examples might include the following: (1) Turn a plastic bottle upside down and watch the water run out, compared to rotating the bottle of water round and round rapidly and not letting the water run out. (2) An old milk jug with a snap-on cap is laid on its side, then a child jumps on it to create a shock wave that sends the cap across the room. This can be aimed at a target if desired. (3) A sheet of paper held horizontally is dropped. Then an identical sheet is crumpled and dropped. Which sheet hit the floor first?

Solution 3.6 A Problem of Waste from the University

A partnership between the City of Newark and the University of Delaware has operated a "U Don't Need It" site for several years. It has kept about 600 t of trash out of the landfill and has helped many poor people, although it is not entirely free of problems. An attendant must meet every incoming vehicle to determine what can be reused or recycled before people throw things into the dumpsters. Many reusable items are in the black plastic bags described as "just trash," including unopened items of food, clocks still operating, china plates and cups, silverware, clothing, and even coins. If the bag rattles, it is opened and bottles/cans are placed in the single-stream recycling system. The site gets more sofas, television sets, and mattresses than it can handle and there is no place to store items over the summer. One professor used his garage for this purpose and made a profit by selling the items to new students. The University of Michigan stores similar items under the football stadium.

Solution 3.7 The Three R's of the Environment

Just appealing to the community probably will not work. Americans usually do not do anything until a crisis occurs. The reusable box is already used by individual companies, such as those selling bread. For general use, it may be hard to balance what is going back with what is coming in. Can you suggest other single-use items that could be reused? This should be easy in our throwaway society. What can be made smaller or the use of it reduced?

Solution 3.8 A Plastic Bottle Opportunity

- Chemical vapor deposition can also put barriers of silicon dioxide on the insides of plastic bottles. This would avoid having to get a license from Mitsubishi. This process could also be used for packaging wine and food.
- An even better method for reducing packaging is to sell nonperishable food in bulk. In this your company could make the plastic bottles. It would be a good idea to offer a discount to get the bottle and the customer back. This system has worked well with bisphenol polycarbonate containers for orange juice and milk. Since bisphenol has been shown to be an endocrine disruptor, your containers could replace it. Leasing the bottle may not be a good idea.

SOLUTIONS TO CHAPTER 4: ENERGY AND THE ENVIRONMENT

Solution 4.1 A Sticky Issue

- See if a careful financial analysis will persuade him.
- Lobby for a state law like the one operational in California.
- Get support from the senior university administration and/or alumni.
- Look for government grants or help from local companies that are trying to make progress towards a sustainable future.
- Persuade the state to require the university to submit and adopt a five-year plan. Cut the university's appropriation if it does not comply.

Solution 4.2 A Nuclear Fantasy

There are already three nuclear reactors across the river in New Jersey. They have had problems of safety and reliability. The cooling water intake kills a large number of fish, which must have a big impact on the fishery in Delaware Bay. The company has refused to put in more cooling towers and has negotiated an agreement with the state to restore some marshes instead.

Even newer is the proposal to build minireactors to be installed underground with a full load of fuel and a system for waste to be kept underground. Refueling might be done every thirty years. Students usually do not like the idea of putting one between the chemistry building and the administration building to reduce transmission losses. The first such reactor is now being built to be installed elsewhere.

Solution 4.3 A Hectic Life

- Your technician might be allowed to use 10% of his time to do his own experiments, if a safety engineer approved them and your division head thought they were in the company's line of business.
- Ask your manager to arrange a transfer of the technician from your laboratory to another one and get a new technician.
- You might quit and look for another job that may be hard to find in a tight job market. You could start your own consulting company if you have enough contacts.
- Would The Lamp Company hire you?

Solution 4.4 What We Know How to Do but Are Not Doing

- Passive solar heating and cooling can save half the energy for operating a building at half the cost of a photovoltaic array. California is the only state that requires it. The initial cost will be 5%–10% higher, but the payback time will be short. Architects know how to do it and are ready to start. The appearance of the building and the neighborhood will be different, but will still look nice.
- District heating and cooling can recover up to 85% of the energy in a fuel. It is used in Europe, but hardly at all in the United States. It also saves more fish and other aquatic organisms than when the waste heat causes river temperatures to rise.
- We know how to put five people into a car and 40 people in a bus. Most commuters drive alone in their cars. Some college students go into debt to finance their own vehicles. We could try living closer to work, ride bikes, and get sidewalks and greenways put in.
- Items used to be returned to the store and reused. Today is the era of single-use, throwaway items. This has made the United States the nation with the most waste per capita. We may have to relearn how to wash dishes.
- There was a time when clothes were dried on an outdoor clothesline. Now, there are even deed restrictions against this.
- Bottled water is a waste of energy, materials, and money. Some cities have prohibited its use.
- There was a time before TV, which can be a real-time waster and contributes to obesity.
- We used to live without plastic grocery bags. You can take your own canvas bag when you go to the store.
- Lawns used to be cut with a reel mower pushed by hand. Now, the ride-on gasoline mower may be a status symbol in the neighborhood. We used to rake leaves with a hand rake, but now we use a leaf blower.
- Ready-to-eat breakfast cereals used to contain no sugar or just a small amount, for example, shredded wheat, puffed wheat, puffed rice, or grape nuts. Now, a cereal labeled "NEW" means more sugar added, even until the cereal is 50% sugar. This has increased dental caries and obesity.
- Finally, we know that we should follow the Institute of Medicine—United States Department of Agriculture recommendations for a healthy diet (no

more than 1500 mg sodium per day, no more than 37 g sugars per day (male, 25 g female), a minimum of saturated and trans fats, and lots of vegetables, fruits, and whole grains).

Solution 4.5 An Episode of Fracking

- Investigate the stories carefully to get the facts. Analyze water samples from receiving rivers.
- See what is put down the well and what comes back up. This might call for a change in the proprietary compositions.
- Try to prevent leaks of methane, which is a stronger greenhouse gas than carbon dioxide.

Solution 4.6 Can More Material Goods Lead to Happiness?

Try removing the tax deduction for mortgages on houses above a certain value. Multiple studies have shown that once a person's basic needs are satisfied, money is no longer a good incentive. A better approach is the recognition of one's ability by one's peers for a job well done by an award. Genuine praise from one's peers also helps.

Even the very wealthy have problems such as worrying about when an unhappy board of directors may replace them, children who are not behaving and such. Money cannot buy love or happiness. A person in a mud hut in a developing nation may be happier than one in a million-dollar mansion.

SOLUTIONS TO CHAPTER 5: ENVIRONMENTAL ECONOMICS
Solution 5.1 To Veto or Not

- This type of monitoring has been hard to implement in many places, both at the state and federal levels. The long-term Framingham Heart Study (www. framinghamheartstudy.org) has uncovered many relations between health and diet. It has shown the value of such monitoring.
- It is typical for states to look to gambling to bring in additional revenue.
- Public funding of campaigns for election would help, but is unpopular.

Solution 5.2 An Issue at Sea

- Build a double-walled tanker.
- Inspect welds by ultrasound when the ship is built.
- Most oil in United States waters comes from motorists who change their own oil and dump it down the drain. Street drains are now being labeled "Drains to River."
- Install easy places for motorists to take used oil. Give them a credit on their next purchase of motor oil for doing this.
- Re-refine the oil for reuse.[10–12]

Solution 5.3 The Precious Plant

- Isolate the active compound (Alnusin III) and determine its structure. Can it be synthesized cheaply and efficiently?
- Investigate if other plants, other parts of *Alnus maritima*, or other species of *Alnus* contain the compound.
- Cultivate the plant (although growth may be slow)/trim hedgerows.
- Put the genes responsible for its synthesis in *Escherichia coli*.
- Develop a plant tissue culture to produce the compound.

This problem is patterned after Taxol (paclitaxil) and the Pacific yew tree. Screening different types of yew trees showed that the compound was present in other species. Trimming the hedgerows of yews in nurseries provided material from which Taxol could be produced. The current method for production is plant tissue culture, which took several years to optimize.

Solution 5.4 Compound X

- Administer Compound X to appropriate test animals. What happens when lower levels are given over a long period of time? It would be unethical to test it on paid volunteers.
- Look for "hot spots" where the residents are consuming much more than the average amounts. Do they have any of the problems caused by large amounts used with test animals?
- Test the compound on various human cell cultures.
- Run blood tests to identify individuals with large amounts. Are they any sicker?
- Put in a large safety factor to protect infants, children, and pregnant women.

Solution 5.5 A Valuable Community Asset

- Hire the outside contractors if your own personnel do not have the time or the expertise to do it.

This happened at the refinery in Delaware City, Delaware, which had a terrible environmental record when operated by Motiva. The plant did shut down later for economic reasons, but has since been restarted with the state's blessing as PBF Energy. It processes 210,000 barrels/day of heavy sour crude oil. Its environmental record has improved.[13]

Solution 5.6 Fast Food Litter

- Charge an extra dollar for takeout items.
- Switch to china plates and cups in fast food restaurants and put in a dishwasher. This will create employment opportunities.
- Enforce a US$100 fine on people who discard fast food litter on the roadsides, although this would be challenging to implement.

- Install cameras on roads near fast food restaurants to photograph the license plate and send a bill to the owner of the car. This is done now for going through a traffic red light.

Solution 5.7 Too Many Cars

Reasonable solutions will be university- and campus-specific. Are there any other possibilities?

Solution 5.8 A Problem with Paper

- Get the college dietician to put even healthier soy products in the dining halls.
- Waterproof the inside of the paper box.
- Charge extra money for takeout orders.
- Use china plates and cups, and clean them in a dishwasher. This will create jobs for college students.

Solution 5.9 Should TOSCA Be Replaced or Revised?

- TOSCA does not require a company to do anything about its toxic emissions. It also assumes that a compound is safe until proven otherwise. Too many companies have hidden under a provision for not reporting proprietary information.
- Environmentalists would like a system more like the European REACH (Registration, Evaluation, Authorization and Restriction of Chemicals) regulation, where the company must show the compound to be safe before entering the market. American companies that market their products in Europe will be subject to REACH.

Companies do not like the use of the precautionary principle, which says that if there is a reasonable doubt one should err on the side of caution. A classic case was the use of tetraethyllead as an antiknock agent, which had to be outlawed when it was shown to be lowering the IQ of children. The producers knew that ethanol would work, but chose tetraethyllead because ethanol could not be patented and tetraethyllead could.

Solution 5.10 A Downward Spiral

- Design a manufacturing process to use less of the element.
- Design an approach to substitute an element that is more abundant.
- Develop ways to recover and recycle the element.
- Look for more ores to exploit.
- Rework landfills to recover metals.
- If possible, obtain the element from seawater. This has been done with uranium, but it costs ten times as much as uranium mined on land.
- Certain nations can corner the market, restrict what is available, and raise the price. China has done this with certain rare earth metals.

- Phosphorus is critical for life, with the main source now being from Morocco. Can we afford to put it in detergents or let it run off lawns or croplands to cause eutrophication of the receiving waters?

SOLUTIONS TO CHAPTER 6: THE GREENING OF SOCIETY

Solution 6.1 Manufacturing Products for a Sustainable Future

- Interface does not sell carpets, it just rents and services them. When new carpets are needed, the company recycles them chemically.
- DuPont became contracted to paint cars for a car manufacturer. Thus, the incentive was to use as little paint as possible to get the job done. This might even involve a new painting technique.
- Electrical utilities are reducing demand for electricity by shifting customers away from times of peak demand, so that expensive auxiliary generators do not have to be put into action.

Solution 6.2 Remedying a Herbal Problem

- Test for effectiveness and safety, bearing in mind that this could cut into your profits.
- Remind the public that many traditional medicines have led to drugs, such as artemisinin for malaria, and aspirin from willow bark.
- Isolate the bioactive substances and use as prescription drugs.
- Clean up your own act![14]

Solution 6.3 Strong Paper Additives?

- Check the patent literature regarding "Strengthon."
- Check to see what has been published in trade journals.
- Cut your price and your profit.
- License their process. Customers like to have more than one source of supply.
- Acquire the company.

Solution 6.4 A Strange New Disease

- The wells were installed, but the water was not tested (at least not for arsenic, which was causing the epidemic).
- Scan the medical literature to see if the disease had been reported before. Taiwan and Chile also have the same problem.
- Hunt for cheap, easy-to-use methods to remove the arsenic. What should be done with the recovered element?[15]

Solution 6.5 "Adhere" versus "SuperGloop"

- Check their patents.
- Have your salesmen get a sample from a customer and analyze it. There are outside companies that specialize in such analyses.

- Check the trade journals for comparable work and the company's advertisements for hints. Could nanotechnology be used in the adhesive?

Solution 6.6 A Water Catastrophe

The problem was due to Cryptosporidium, a protozoan that is not killed by chlorine. It may have entered into the water supply from the pastures where the cows are kept due to wet weather. This happened in Milwaukee, Wisconsin several years ago. The city now uses ozone to disinfect the water and a small amount of chlorine to keep it clean in the pipes.

Solution 6.7 Avoiding an Expensive Recoat

- Check to see if they have any patents filed and if so, whether or not they infringe your patents.
- Send the company a legal warning letter.
- Advertise the way(s) in which your product is superior.
- See whether the zoning actually allows chemical manufacture. See what trucks go in that are bringing raw materials.
- Check whether the company is violating any state air and water emissions laws.
- Verify whether their customers see any advantages to the product beyond a reduced price.
- Purchase their company.

Solution 6.8 Battling the Business Center

- Check to see whether the Japanese company has patented their process.
- Search for the toll manufacturer in China.
- Perform research to establish whether the brown color can be removed with activated carbon or via another process.

Solution 6.9 Your Viewpoint on Some Green Chemistry and Sustainability Issues

1. The recent problems at the Fukushima plant in Japan make it questionable. Conservation of energy is the best way to operate.
2. "Natural" is a meaningless term not recognized by the USDA.
3. You are taking a chance with your health by using herbal supplements. Most of them do not work and may interfere with prescribed medicines. They do not contain standard amounts.
4. The FDA is here to protect us.
5. Probably not. That does not mean that some laws cannot be simplified or improved.
6. Industry laboratories are usually safer.
7. Dermatologists say that there is no such thing as a healthy tan. The tan indicates injury to the skin.
8. Accidents are unnecessary. Be careful and try not to have any.

9. The oil refinery at Delaware City has been restarted with state help and has been trying to clean up its act.

10. Safety inspections can help if they are done wholeheartedly.

11. The Institute of Medicine USDA recommendations are for a maximum of 1500 mg sodium per day. The average American eats twice this much every day. Too much salt can raise blood pressure and cause a stroke.

12. Cell phones are very popular so there have to be cell phone towers. Some have been designed to look like redwood trees.

REFERENCES

1. L. Greer and C. Van Löben Sels, *Environ. Sci. Technol.*, 1997, *31*, 418A.
2. B.M. Jenssen, *Environ. Health Perspect.*, 2006, *114 (Suppl. 1)*, 76.
3. R.S. Drago, J.A. Dias and T.O. Maier, *J. Am. Chem. Soc.*, 1997, *119*, 7702.
4. S. Biz and M.L. Ocelli, *Catal. Rev. Sci. Eng.*, 1998, *40*, 329.
5. A. Zecchina, C. Lamberti and S. Bordiga, *Catal. Today*, 1998, *41*, 169.
6. W.J. Darby, *Chem. Eng. News*, Oct. 1, 1962, 60.
7. R.P. Swatloski, S.K. Spear, J.D. Holbrey and R.D. Rogers, *J. Am. Chem. Soc.*, 2002, *124*, 4974.
8. R.G. Hicks, Asphalt Rubber Design and Construction Guidelines, which can be located via Google: "asphalt rubber design and construction guidelines."
9. B. Hileman, *Chem. Eng. News*, July 1, 2002, 15.
10. J. Bohannon, X. Bosch and J. Withgott, *Science*, 2002, *298*, 1695.
11. J. Bohannon and X. Bosch, *Science*, 2002, *299*, 490.
12. P. Serret, X.A. Álvarez-Salgado, A Bode et al. *Science*, 2002, *299*, 511.
13. J-F. Tremblay, *Chem. Eng. News*, Aug. 8, 2011, 11a.
14. M. Angell and J.P. Kassirer, *N. Engl. J. Med.*, 1998, *339*, 839.
15. A.H. Smith, E.O. Lingas and M. Rahman, *Bull. World Health Organ.*, 2000, *78*, 1093.

Index